Energetics
of
Secretion
Responses

Volume II

Editor

Jan-Willem N. Akkerman, Ph.D.
Director
Laboratory of Haematology
University Hospital Utrecht
Utrecht, The Netherlands

CRC Press
Taylor & Francis Group
Boca Raton London New York

CRC Press is an imprint of the
Taylor & Francis Group, an **Informa** business

THE EDITOR

Jan-Willem N. Akkerman, Ph.D. is the Director of the Laboratory of Haematology in the Department of Haematology at the University of Utrecht, The Netherlands. He received his Ph.D. degree in 1975 on a thesis entitled ''Human Platelet Glycolysis'' and was a visiting staff member at the Thrombosis Research Center at Temple University, Philadelphia,Pa. in 1978.

His research efforts are in the field of thrombosis and atherosclerosis, with special emphasis on the role of blood platelets and the mechanisms that make these cells responsive to platelet-activating agents. His current research interests are the energetics of platelet functions, mechanisms for signal transduction, and abnormal platelet behavior in patients with bleeding disorders or hyperaggregability.

CONTRIBUTORS

Volume II

Jan-Willem Akkerman, Ph.D.
Director
Laboratory of Haematology
University Hospital Utrecht
Utrecht, The Netherlands

Katsuhiko Ase
Researcher
Department of Biochemistry
Kobe University School of Medicine
Kobe, Japan

Keith W. Brocklehurst, Ph.D.
Visiting Associate
Laboratory of Cell Biology and Genetics
National Institute of Diabetes, and
  Digestive and Kidney Diseases
National Institutes of Health
Bethesda, Maryland

James L. Daniel, Ph.D.
Associate Professor
Department of Pharmacology
Temple University
Philadelphia, Pennsylvania

Masayoshi Go, M.D.
Graduate Student
Department of Biochemistry
Kobe University School of Medicine
Kobe, Japan

Tadashi Hashimoto, M.D.
Graduate Student
Department of Biochemistry
Kobe University School of Medicine
Kobe, Japan

Ushio Kikkawa, M.D., Ph.D.
Lecturer
Department of Biochemistry
Kobe University School of Medicine
Kobe, Japan

Sigurd Lenzen, M.D.
Professor
Institute of Pharmacology and Toxicology
University of Göttingen
Göttingen, West Germany

Willy J. Malaisse, M.D., Ph.D.
Professor
Laboratory for Experimental Medicine
Brussels Free University
Brussels, Belgium

Yasutomi Nishizuka, M.D., Ph.D.
Professor and Chairman
Department of Biochemistry
Kobe University School of Medicine
Kobe, Japan

Hideaki Nomura, M.D.
Graduate Student
Department of Biochemistry
Kobe University School of Medicine
Kobe, Japan

Kohji Ogita, M.D.
Graduate Student
Department of Biochemistry
Kobe University School of Medicine
Kobe, Japan

Uwe Panten, M.D.
Professor
Institute of Pharmacology and Toxicology
University of Göttingen
Göttingen, West Germany

Harvey B. Pollard, M.D., Ph.D.
Laboratory Chief
Laboratory of Cell Biology and Genetics
National Institute of Diabetes, and
  Digestive and Kidney Diseases
National Institutes of Health
Bethesda, Maryland

Abdullah Sener, Ph.D.
Lecturer
Laboratory of Experimental Medicine
Brussels Free University
Brussels, Belgium

Winfried Siffert, Dr.med.
Post Doctoral Fellow
Institute for Physiology
Ruhr University
Bochum, West Germany

Jan Wilschut, Ph.D.
Associate Professor
Department of Physiological Chemistry
University of Groningen
Groningen, The Netherlands

TABLE OF CONTENTS

Volume I

TABLE OF CONTENTS

Volume II

Energy-Dependent Steps in Signal Processing

Chapter 9

CALCIUM MOBILIZATION

W. Siffert

TABLE OF CONTENTS

I. INTRODUCTION

It is now generally accepted that translocation of calcium ions forms an essential step in the sequence that couples receptor activation to the mechanisms that execute exocytosis. Most secretagogues induce a rapid rise in cytosolic free Ca^{2+}, partly as a result of Ca^{2+} influx from the extracellular medium and partly as a result of Ca^{2+} liberation from intracellular storage sites. This chapter summarizes some aspects of the role of Ca^{2+} in secretion responses and its sensitivity to alterations in intracellular pH, cAMP, and cGMP concentrations.

II. ROLE OF CALCIUM

A. Ca^{2+} Homeostasis in Unstimulated Cells

Unstimulated cells maintain a low cytosolic free Ca^{2+} concentration ($[Ca^{2+} i]$) in spite of a large, inward directed Ca^{2+} gradient across the plasma membrane. Both the plasma membrane and intracellular membranes contribute to Ca^{2+} homeostasis in unactivated cells.[1-3] The plasma membrane itself is rather impermeable for Ca^{2+} [1] and Ca^{2+} ions that enter the cell due to diffusion are extruded by several energy requiring mechanisms such as a Ca^{2+}-H^+-ATPase, a calcium pump, and Na^+-Ca^{2+} exchange, the latter being supported by the Na^+-K^+-ATPase.[1-6] Another system that removes Ca^{2+} from the cytosol of secretory cells is the endoplasmic reticulum. Internal membranes derived from human platelets actively accumulate Ca^{2+} in the presence of ATP[7,8] and a similar Ca^{2+}-sequestrating system has been described in permeabilized human neutrophils.[9] Although the total energy cost to the cell for maintaining Ca^{2+} homeostasis is not exactly known, Rasmussen and Barret have estimated an energy requirement of about 20 μmol ATP min^{-1} kg^{-1} cell H_2O which equals about 0.2% of the basal energy consumption of the cell.[2]

B. Ca^{2+} Mobilization in Activated Cells

The development of fluorescent, intracellularly trappable Ca^{2+} indicators such as quin 2,[10] fura 2, and indo 1[11] has encouraged many investigators to study Ca^{2+} translocation in activated cells.

Among the secretory cells that respond to stimulation with an increase in $[Ca^2 i^+]$ are platelets,[12-15] neutrophils,[16,17] adrenal chromaffin cells,[18] macrophages,[19] pancreatic acinar cells,[20] parathyroid cells,[21] and many others. The increase in $[Ca^2 i^+]$ following receptor occupancy is partially due to the release of Ca^{2+} from internal, nonmitochondrial stores, since substantial increases in $[Ca^2 i^+]$ can be observed after reducing the external Ca^{2+} concentration below $10^{-9} M$ by addition of EGTA to the incubation medium.[12-15] The putative second messenger mediating Ca^{2+} liberation from the endoplasmic reticulum is inositol 1,4,5-trisphosphate (IP_3), a product of the phosphodiesteratic cleavage of phosphatidylinositol 4,5-bisphosphate.[22,23] A direct effect of IP_3 on Ca^{2+} mobilization in permeabilized cells or isolated internal membrane preparations has actually been demonstrated in neutrophils,[9] platelets,[24,25] pancreatic islets,[26] and pancreatic acinar cells.[20,27] Recent interest has focused on the question of whether GTP or a GTP-binding protein might play a role in IP_3-mediated Ca^{2+} release.[28] A role of GTP in the mobilization of Ca^{2+} from guinea pig parotid gland endoplasmic reticulum,[29] from a rat liver microsomal fraction,[30] and from endoplasmic reticulum derived from a neuroblastoma cell line[31] has recently been demonstrated. Such a GTP-dependent mechanism of Ca^{2+} release would provide an additional link between cell activation and energy metabolism.[32] However, the major mechanism by which $[Ca^2 i^+]$ is increased after receptor stimulation is influx of Ca^{2+} across the plasma membrane, but the molecular events underlying this process are poorly understood. Possible mechanisms might involve voltage-dependent Ca^{2+} channels, receptor-operated channels, and lipid mediators of calcium mobilization, which might act as "Ca^{2+} ionophores".[2,4,33,34]

C. The Role of Ca^{2+} in Secretion

Our insight in the role of intracellular second messengers in cellular signal transduction has been greatly deepened by the development of techniques that selectively permeabilize the plasma membrane, but leave internal membranes unaffected. Such techniques include electropermeabilization, permeabilization by detergents (saponin or digitonin), or by incubation of cells with Sendai virus. These methods have made it possible to introduce small molecules into the cytosol, e.g. Ca^{2+} ions, nucleotides etc., in order to study their role in secretion.[36]

When permeabilized platelets are exposed to increasing concentrations of Ca^{2+}, secretion is initiated even in the absence of an agonist[37,38] and almost identical results are obtained in pancreatic acini,[39] adrenal medullary cells, and pancreatic islets.[37] The ability of Ca^{2+} to trigger a secretory response depends on the presence of MgATP^{2-} in the medium. ATP analogs which are not hydrolyzed are ineffective in promoting Ca^{2+}-dependent secretion, suggesting that ATP-dependent phosphorylation or energy liberation from ATP hydrolysis is an absolute requirement.[36] Although in permeabilized human platelets Ca^{2+}-dependent secretion is accompanied by protein phosphorylation, a causal relationship between protein phosphorylation and secretion has not yet been established.[40,41] Furthermore, almost a similar degree of protein phosphorylation can be observed in the absence of measurable secretion.[42,43] Therefore, it may well be that energy provision by ATP hydrolysis is required for other processes and Knight and Scrutton have suggested, that this energy may be needed "to bring the secretory vesicle and the plasma membrane into contact with each other".[36]

Further evidence for a crucial role of Ca^{2+} in secretion has been derived from experiments in which cells were loaded with high concentrations of the fluorescent intracellular Ca^{2+} indicator quin 2. Under conditions where the intracellular dye concentration approaches values of about 5 mM, quin 2 chelates [Ca^2 i$^+$] and acts as a Ca^{2+} buffer. Loading human platelets with high quin 2 concentrations significantly reduces both the stimulus-induced increase in [Ca^2 i$^+$] as well as secretion and aggregation.[44,45] Inhibitory effects of quin 2 were also reported on the secretion of growth hormone and prolactin from anterior pituitary cells,[46] the chemotactic peptide-induced granule exocytosis from neutrophils,[47] and the acetylcholine-stimulated secretion of catecholamines from adrenal chromaffin cells.[18]

Taken together, it appears that Ca^{2+} plays a central role in the cascade of events that leads to secretion and that for this role metabolic energy is required.

D. Is there Ca^{2+}-Independent Secretion?

New impulses for research on stimulus-response coupling mechanisms were obtained from the observation, that the binding of an agonist to its specific membrane receptor induces a rapid turnover of certain phospholipids. It is now generally accepted that stimulation of cells by hormones or other agonists activates phospholipase C, an enzyme which hydrolyzes phosphatidylinositol 4,5-bisphosphate to IP$_3$ and 1,2 diacylglycerol (DG). IP$_3$ releases Ca^{2+} from internal membranes, whereas DG activates protein kinase C (C-kinase), which in turn phosphorylates several proteins which are apparently important for secretion.[22,23,48-50] After the discovery that tumor promoting phorbol esters, such as 12-O-tetradecanoylphorbol-13-acetate (TPA) initiate secretion via activation of the C-kinase thereby mimicking the action of DG, TPA became a popular tool to investigate the role of this enzyme in secretion. Activators of the C kinase and Ca^{2+}-ionophores induced an incomplete secretion response, but in combination, these agents completely mimicked the effects of natural agonists. Subsequently, a model for the activation mechanism of secretion was proposed which postulated an early bifurcation in the signal transduction mechanism with a C-kinase pathway and a Ca^{2+}-dependent route. Both pathways would act synergistically at a later stage of cell activation thereby inducing the complete secretion response.[48]

In 1983 Rink et al. demonstrated that activators of the C-kinase such as TPA or 1-oleoyl-

2-acetylglycerol (OAG) stimulated secretion from human platelets without raising cytoplasmic free Ca^{2+} (as judged from the quin 2 fluorescence) both in the absence and presence of external Ca^{2+}.[51] Collagen, an important physiological activator of platelet reactions, also induced the secretion of granule contents without increasing $[Ca^{2+}]$, despite a normal formation of IP_3, i.e., the signal for intracellular Ca^{2+} mobilization.[51,52] However, other investigators presented evidence that OAG does not induce a significant secretory response by itself but enhances secretion evoked by natural agonists.[42,43,53] Ware et al. reinvestigated the effect of OAG and TPA on $[Ca^2 i^+]$ using aequorin-loaded platelets and found that both agonists significantly increased the cytosolic free Ca^{2+} concentration.[54] They proposed that aequorin and quin 2 reflected different aspects of Ca^{2+} homeostasis and concluded that "the absence of quin 2 signal may not be sufficient to establish that a metabolic pathway is Ca^{2+}-independent". Similar conclusions were drawn by Knight and Scrutton.[38] Although TPA and OAG lowered the apparent Ca^{2+} requirement of serotonin secretion in permeabilized platelets, the effectiveness of both C-kinase activators was reduced considerably by increasing the Ca^{2+}-buffering capacity of the medium.

Contradictory results regarding the role of Ca^{2+} and the C-kinase in secretion were also obtained in neutrophils. TPA stimulated lysozyme secretion from intact neutrophils without increasing $[Ca^2 i^+]$,[55] but inhibited Ca^{2+}-induced secretion of β-glucuronidase in permeabilized cells.[56]

Thus, apart from the technical difficulties of measuring $[Ca^{2+}i]$, other, still unidentified factors may modulate the relationship between $[Ca^{2+}i]$ and the secretion response.[36]

III. ROLE OF CYTOPLASMIC pH

A. pH Homeostasis in Resting Cells

Unstimulated cells maintain a cytoplasmic pH (pH_i) of about 7.0 to 7.2. This is generally lower than the extracellular pH (about 7.4), but distinctly higher as one would expect if H^+ ions were in electrochemical equilibrium across the plasma membrane. Under these conditions pH_i would approach values of 6.3 to 6.5, because of a larger interior-negative membrane potential.[57,58] It is quite clear that regulation of pH_i requires an energy-dependent mechanism for acid extrusion. The plasma membranes of many cells contain a transport system that mediates an electroneutral transmembrane exchange of external Na^+ against internal H^+, and this transport system is commonly referred to as the "plasma membrane Na^+/H^+ exchange".[57-59] The inwardly directed Na^+ gradient provides the driving force for uphill H^+ extrusion. It has been suggested that Na^+/H^+ antiport activity does not directly depend on metabolic energy, despite the fact that the maintenance of the Na^+ gradient across the plasma membrane by active extrusion of Na^+ via the Na^+-K^+-ATPase system requires energy in the form of ATP.[57,58] A recent report, however, indicates that ATP depletion of cells may partially inhibit the activity of the Na^+/H^+ antiport.[60] Whereas the cytosolic pH of secretory cells is very efficiently maintained within the range of 7.0 to 7.2, other cellular compartments have distinctly different pH values. Secretory granules of mast cells,[61] platelets,[62,63] anterior pituitary,[64] and adrenal chromaffin cells[65] maintain an acidic internal pH which is approximately 1 pH unit lower than the cytoplasmic pH. An H^+-translocating ATPase is responsible for the generation of this electrochemical proton gradient and the resulting pH is maintained by the low H^+ permeability of the granule membrane, in combination with the high internal buffer capacity. The transmembrane proton gradient across these secretory granules is the driving force for the uptake and storage of biogenic amines.[62-66]

B. Stimulus-Induced Changes in Cytoplasmic pH

Many cells respond to stimulation with an increase in pH_i via activation of the Na^+/H^+

antiport.[68-71] Recent investigations have suggested that the antiport is stimulated by a C-kinase-dependent pathway, because the action of natural agonists could be mimicked by activators of protein kinase C (e.g. TPA and OAG[68,69]). It has also been demonstrated that the phorbol ester-induced activation of Na^+/H^+ exchange depends on metabolic energy since ATP depletion of cells completely suppressed the TPA-induced cytoplasmic alkalinization.[72]

C. Role of pH$_i$ Changes in Secretion

Among the secretory cells that respond to stimulation with an increase in pH$_i$ are neutrophils,[70,73,74] platelets,[75,76] and pancreatic acini.[77] The following discussion is focused on Na^+/H^+ exchange in platelet activation but probably also applies for other secretory cells.

In 1978 Horne and Simons reported that amiloride, a drug which is commonly used to block Na^+/H^+ exchange, inhibited thrombin-induced platelet aggregation as well as the secretion of serotonin.[78] A quite similar inhibition of epinephrine-, ADP-, and thrombin-induced serotonin secretion was observed when platelets were activated in Na^+-free media,[79,80] which made Limbird and colleagues postulate "a link between the presence of extraplatelet Na^+ and the operation of platelet function mediated by the cyclooxygenase pathway". Their evidence was based on the fact that removal of extracellular Na^+ or inhibition of the cyclooxygenase pathway by indomethacin resulted in an identical inhibition of serotonin secretion. They demonstrated that inhibition of Na^+/H^+ exchange blocked the stimulus-provoked arachidonic acid release in human platelets.[81] In a recent study,[82] the role of Na^+/H^+ exchange in phospholipase A2 activation was investigated in detail. This Ca^{2+}-dependent enzyme plays a central role in stimulus-response coupling in platelets[33] and its activation optimum lies in the alkaline pH range.[83] Perturbants of Na^+/H^+ exchange blocked the activation of phospholipase A_2 induced by epinephrine, ADP, and low concentrations of thrombin. No effect of Na^+/H^+ exchange on enzyme activation was observed if $[Ca^{2+}_i]$ was increased by the Ca^{2+} ionophore A23187, although cytoplasmic alkalinization decreased the threshold concentration of A23187 required to elicit enzyme activation. The authors suggested a synergistic action between Ca^{2+} mobilization and cytoplasmic alkalinization in the activation of phospholipase A_2 and subsequent induction of the secretory responses.[82]

Our own recent experiments have shown that activation of Na^+/H^+ exchange is a prerequisite for Ca^{2+} mobilization in human platelets.[84] Blocking Na^+/H^+ exchange by removal of extracellular Na^+ or by an amiloride analog inhibited the thrombin-induced increase in $[Ca^{2+}_i]$ in quin 2-loaded platelets, whereas artifical cytoplasmic alkalinization drastically enhanced the stimulus-induced Ca^{2+} mobilization. Increasing the pH$_i$ without adding an agonist did not change $[Ca^{2+}_i]$ which indicates that changes in pH$_i$ alone are incapable of initiating a cellular response. Additional support for these findings is found in work from Leven et al.[85] who demonstrated that cytoplasmic alkalinization induced platelet shape change but did not lead to a substantial secretion of serotonin. It thus appears that a stimulus-induced increase in pH$_i$ forms an essential step in cell activation by facilitating Ca^{2+} mobilization. There is good evidence that one of the major targets of cytoplasmic alkalinization is IP$_3$-mediated Ca^{2+} release from internal storage sites which greatly depends on pH in the range of 6.8 to 7.4.[86]

It can be concluded that stimulus-induced changes in pH$_i$ play a role in the early reactions following receptor stimulation and particularly affect Ca^{2+} mobilization. It should be noted, however, that other pH-dependent reactions may be involved in the sequence of events leading to secretion and possible candidates are the pH-dependent interaction of Ca^{2+} with calmodulin and the production of metabolic energy in glycolysis.[58,67]

IV. ROLE OF cAMP

The role of cAMP and Ca^{2+} in cellular signal transduction has been extensively reviewed[1,2,4,33] and this report will only briefly address some recent findings regarding the role of this nucleotide in secretion. Again, the human platelet shall be considered as a model for other secretory cells.

In general, it appears that cAMP plays an inhibitory role in stimulus response coupling in platelets and acts mainly at rather early stages of cell activation. Pretreatment of platelets with stimulators of adenylate cyclase, e.g., prostaglandins E_1 and D_2 or forskolin, inhibited the stimulus-induced rise in $[Ca^{2+} i]$.[33,87,88] If the cells were stimulated by thrombin and activators of the adenylate cyclase were added after $[Ca^2 i^+]$ had reached its maximum, the resultant increase in cAMP levels led to an accelerated removal of Ca^{2+} from the cytosol and even reversed the phosphorylation of the myosin light chain and the 47-kdalton protein.[87] Part of these results can be explained by a stimulatory effect of cAMP on Ca^{2+} accumulation into intracellular Ca^{2+} stores. It has been demonstrated that calcium uptake into Ca^{2+}-accumulating vesicles derived from human platelets is stimulated by cAMP and it was suggested that a cAMP-dependent protein kinase would play a role.[89,90] However, cAMP inhibits the formation of second messengers as illustrated by the effect of prostaglandin I_2 which inhibits the formation of inositol phosphates in thrombin-stimulated platelets.[91] The view that cAMP inhibits stimulus-response coupling by inhibiting second messenger formation found support in studies by Knight and Scrutton[37] who measured Ca^{2+}-induced serotonin secretion from permeabilized platelets. Increasing the Ca^{2+} concentration of the suspending medium led to a substantial secretion of serotonin, which was unaffected by cAMP. The presence of thrombin or OAG in the suspending medium enhanced the Ca^{2+} sensitivity of the secretory mechanism and this effect could be prevented by addition of cAMP.[37,38] A similar system is likely to function in human neutrophils. Della Bianca et al. have recently investigated the effect of cAMP on phosphoinositide turnover in neutrophils stimulated by the chemotactic peptide fMet-Leu-Phe.[92] Increasing the intracellular cAMP levels by preincubation with dibutyryl cAMP, theophylline, or prostaglandin E_1 inhibited the chemotactic peptide-induced breakdown of phosphoinositides as well as phosphatidic acid formation. The generation of inositol phosphates was also markedly reduced. It has been proposed that cAMP inhibits the activity of phospholipase C.[91] At present, it remains uncertain whether this nucleotide inhibits this enzyme directly or indirectly, for instance by acting on a regulatory unit (e.g. G proteins).

V. ROLE OF cGMP

cGMP is formed from GTP through the action of guanylate cyclase. The interest in the role of this cyclic nucleotide in stimulus secretion coupling is now rapidly growing.

Initially contradictory effects of cGMP were reported. Some evidence suggested that cGMP might be an inhibitor of cell activation. Haslam et al.[93] reported that stimulation of platelets by collagen increased the cGMP levels. However, if platelets were pretreated with nitroprusside, which resulted in a two-to sixfold increase in cGMP levels, collagen-induced secretion and aggregation were inhibited. Sodium nitroprusside and the membrane-permeable 8-bromo-cGMP induced the phosphorylation of two polypeptides in intact platelets suggesting the presence of a cGMP-dependent protein kinase.[93] MacIntyre et al.[94] have recently investigated the effects of cGMP on Ca^{2+} mobilization in response to platelet activating factor and proposed that cGMP inhibits the activation of phospholipase C and, thereby, the formation of the intracellular messengers required to increase $[Ca^{2+} i^+]$. Only recently a cGMP-dependent protein kinase and cGMP-dependent phosphorylation have been described and both may be activated by agents which increase the level of cGMP.[95] Quite different results have been obtained in permeabilized platelets. The sensitivity of the Ca^{2+}-induced serotonin secretion in the presence of thrombin or OAG was markedly enhanced by micromolar concentrations of cGMP.[37,38] However, cGMP had no effect when secretion was induced by increasing the free Ca^{2+} concentration in the absence of an agonist.

Interestingly, new impulses for research on the role of cGMP come from an area that is quite different from the field of stimulus-secretion coupling. The discovery of a new hormone, atrial natriuretic factor (ANF), which is released from the heart, has attracted much attention. ANF induces a variety of biological effects including diuresis, smooth muscle relaxation, and inhibition of renin secretion.[96,97] The relaxing effect of ANF on smooth muscle was reported to result from inhibition of intracellular Ca^{2+} release[98] and elevation of intracellular cGMP levels.[99,100]

The effect of ANF appears to be mediated via cGMP-dependent protein phosphorylation.[101] Interestingly, ANF inhibits secretion of aldosteron from zona glomerulosa cells by increasing cGMP levels.[102] If these results reflect a general property of secretory cells, ANF may deliver a valuable contribution in clarifying the role of cGMP in stimulus-secretion coupling.

VI. CONCLUDING REMARKS

It appears that an increase in the cytosolic free Ca^{2+} concentration is one of the most crucial steps in the initiation of a secretory response. In platelets and several other secretory cells cyclic nucleotides provide a negative signal in the sequence of events and probably act by preventing the formation of second messengers required for Ca^{2+} mobilization as well as by increasing the sequestration into intracellular storage sites. Stimulus-induced changes in cytoplasmic pH seem to play a major role in cellular signal transduction. However, it remains to be investigated whether cytoplasmic alkalinization provides a general, rather unspecific stimulating effect or acts on specific cellular targets.

ACKNOWLEDGMENTS

I am deeply indebted to Dr. J. W. N. Akkerman for stimulating discussions and helpful comments during the preparation of this article and to Maeyken Hoeneveld and Annemiek Beyer for typing the manuscript.

Work from the author's laboratory was supported by a grant from the Deutsche Forschungsgemeinschaft (no. Sche 46-5/1) and carried out in close cooperation with Dr. J. W. N. Akkerman. Further support was obtained from the Beauttragle für die Pflege und Forderung der Beziehungen Zwischen den Hochschulen des Landes Nordrhein-Westfalen und des Königreiches der Niederlande und des Königreiches Belgien.

REFERENCES

1. **Borle, A. B.,** Control, modulation and regulation of cell calcium, *Rev. Physiol. Biochem. Pharmacol.,* 90, 13, 1981.
2. **Rasmussen, H. and Barrett, P. Q.,** Calcium messengers systems: an integrated view, *Physiol. Rev.,* 64, 938, 1984.
3. **Schatzmann, H. J.,** Calcium extrusion across the plasma membrane by the calcium-pump and the Ca^{2+}-Na^+ exchange system, in *Calcium and Cell Physiology,* Marmé, D., Ed., Springer-Verlag, Berlin, 1985, 18.
4. **Rasmussen, H., Zawalichi, W., and Kojima, I.,** Ca^{2+} and c-AMP in the regulation of cell function, in *Calcium and Cell Physiology,* Marmé, D. Ed., Springer-Verlag, Berlin, 1985, 1.
5. **Niggli, V., Sigel, E., and Carafoli, E.,** The purified Ca^{2+} pump of human erythrocyte membranes catalyzes an electroneutral Ca^{2+}-H^+ exchange in reconstituted liposomal systems, *J. Biol. Chem.,* 257, 2350, 1982.
6. **Blaustein, M. P.,** The interrelationship between sodium and calcium fluxes across cell membranes, *Rev. Physiol. Biochem. Pharmacol.,* 70, 33, 1974.
7. **Käser-Glanzmann, R., Jakabova, M., George, J. N., and Lüscher, E. F.,** Further characterization of calcium-accumulating vesicles from human blood platelets, *Biochim. Biophys. Acta,* 512, 1, 1978.

8. **Adunyah, S. E. and Dean, W. L.,** Ca^{2+} transport in human platelet membranes, *J. Biol. Chem.,* 261, 3122, 1986.

9. **Prentki, M., Wollheim, C. B., and Lew, P. D.,** Ca^{2+} homeostasis in permeabilized human neutrophils, *J. Biol. Chem.,* 259, 13777, 1984.

10. **Tsien, R. Y., Pozzan, I., and Rink, T. J.,** Calcium homeostasis in intact lymphocytes: cytoplasmic free calcium monitored with a new, intracellularly trapped fluorescent indicator, *J. Cell Biol.,* 94, 325, 1982.

11. **Grienkiewicz, G., Poenie, M., and Tsien, R. Y.,** A new generation of Ca^{2+} indicators with greatly improved fluorescence properties, *J. Biol. Chem.,* 260, 3440, 1985.

12. **Rink, T. J., Smith, S. W., and Tsien, R. Y.,** Cytoplasmic free Ca^{2+} in human platelets: Ca^{2+} thresholds and Ca^{2+}-independent activation for shape change and secretion, *FEBS Lett.,* 148, 21, 1982.

13. **Hallam, T. J., Sanchez, A., and Rink, T. J.,** Stimulus-response coupling in human platelets. Changes evoked by platelet-activating factor in cytoplasmic free calcium monitored with the fluorescent calcium indicator quin 2, *Biochem. J.,* 218, 819, 1984.

14. **Hallam, T. J. and Rink, T. J.,** Responses to adenosine diphosphate in human platelets loaded with the fluorescent calcium indicator quin 2, *J. Physiol. (London),* 368, 131, 1985.

15. **Erne, P. and Pletscher, A.,** Rapid intracellular release of calcium in human platelets by stimulation of $5\text{-}HT_2$-receptors, *Br. J. Pharmacol.,* 84, 545, 1985.

16. **Korchak, H. M., Vienne, K., Rutherford, L. E., Wilkenfeld, C., Finkelstein, M. C., and Weissmann, G.,** Stimulus response coupling in the human neutrophil. II. Temporal analysis of changes in cytosolic calcium and calcium efflux, *J. Biol. Chem.,* 259, 4076, 1984.

17. **Sjlar, L. A. and Oades, Z. G.,** Signal transduction and ligand-receptor dynamics in the neutrophil, *J. Biol. Chem.,* 260, 11468, 1985.

18. **Kao, L. S. and Schneider, A. S.,** Calcium mobilization and catecholamine secretion in adrenal chromaffin cells, *J. Biol. Chem.,* 261, 4881, 1986.

19. **Conrad, G. W. and Rink, T. J.,** Platelet activating factor raises intracellular calcium ion concentration in macrophages, *J. Cell Biol.,* 103, 439, 1986.

20. **Streb, H., Heslop, J. P., Irvine, R. F., Schulz, I., and Berridge, M. J.,** Relationship between secretagogue-induced Ca^{2+} release and inositol polyphosphate production in permeabilized pancreatic acinar cells, *J. Biol. Chem.,* 260, 7309, 1985.

21. **Nemeth, E. F., Wallace, J., and Scarpa, A.,** Stimulus-secretion coupling in bovine parathyroid cells, *J. Biol. Chem.,* 261, 2668, 1986.

22. **Berridge, M. J.,** Inositol trisphosphate and diacylglycerol as second messengers, *Biochem. J.,* 220, 345, 1984.

23. **Berridge, M. J., and Irvine, R. F.,** Inositol trisphosphate, a novel second messenger in cellular signal transduction, *Nature (London),* 312, 315, 1984.

24. **Brass, L. F. and Joseph, S. K.,** A role for inositol trisphosphate in intracellular Ca^{2+} mobilization and granule secretion in platelets, *J. Biol. Chem.,* 260, 15172, 1985.

25. **Authi, K. S. and Crawford, N.,** Inositol 1,4,5 trisphosphate-induced release of sequestered Ca^{2+} from highly purified human platelet intracellular membranes, *Biochem. J.,* 230, 247, 1985.

26. **Wolf, B. A., Comens, P. G., Ackermann, K. E., Sherman, W. R., and McDaniel, M. L.,** The digitonin-permeabilized pancreatic islet model. Effect of myo-inositol 1,4,5-trisphosphate on Ca^{2+} mobilization, *Biochem. J.,* 227, 965, 1985.

27. **Streb, H., Irvine, R. F., Berridge, M. J., and Schulz, I.,** Release of calcium from a non-mitochondrial intracellular store in pancreatic acinar cells by inositol-1,4,5-trisphosphate, *Nature (London),* 306, 67, 1983.

28. **Baker, P. F.,** Intracellular signalling. GTP and calcium release, *Nature (London),* 320, 395, 1986.

29. **Henne, V. and Söling, H. D.,** Guanosine 5'-triphosphate releases calcium from rat liver and guinea pig parotid gland endoplasmic reticulum independently of inositol 1,4,5-trisphosphate, *FEBS Lett.,* 202, 267, 1986.

30. **Dawson, A. P., Comerford, J. G., and Fulton, D. V.,** The effect of GTP on inositol 1,4,5-trisphosphate-stimulated Ca^{2+} efflux from a rat liver microsomal fraction, *Biochem. J.,* 234, 311, 1986.

31. **Gill, D. L., Ueda, T., Chueh, S. H., and Noel, M. W.,** Ca^{2+} release from endoplasmic reticulum is mediated by a guanine nucleotide regulatory mechanism, *Nature (London),* 320, 461, 1986.

32. **Alberts, B., Bray, D., Lewis, J., Raff, M., Roberts, K., and Watson, J. D.,** *Molecular Biology of the Cell,* Garland Publ., New York, 1983, 72.

33. **Feinstein, M. B., Halenda, S. P., and Zovoico, G. B.,** Calcium and platelet function, in *Calcium and Cell Physiology,* Marmé, D., Ed., Springer-Verlag, Berlin, 1985, 345.

34. **Malaisse, W. J., Lebrum, P., and Herchulez, A.,** Calcium regulation of insulin release, in *Calcium and Cell Physiology,* Marmé, D., Ed., Springer-Verlag, Berlin, 1985, 298.

35. **Peachell, P. T. and Pearce, F. L.,** Calcium regulation of histamine secretion from mast cells, in *Calcium and Cell Physiology,* Marmé, D., Ed., Springer-Verlag, Berlin, 1985, 311.

36. **Knight, D. E. and Scrutton, M. C.,** Gaining access to the cytosol: the technique and some applications of electropermeabilization, *Biochem. J.,* 234, 497, 1986.

37. **Knight, D. E. and Scrutton, M. C.**, Cyclic nucleotides control system which regulates Ca^{2+} sensitivity of platelet secretion, *Nature (London)*, 309, 66, 1984.

38. **Knight, D. E. and Scrutton, M. C.**, The relationship between intracellular second messengers and platelet secretion, *Biochem. Soc. Trans.*, 12, 969, 1984.

39. **Kimura, T., Imura, K., Eckhardt, L., and Schulz, I.**, Ca^{2+}-, phorbolester-, and cAMP-stimulated enzyme secretion from permeabilized rat pancreatic acini, *Am. J. Physiol.*, 250, G698, 1986.

40. **Knight, D. E., Niggli, V., and Scrutton, M. C.**, Thrombin and activators of protein kinase C modulate secretory responses of permeabilized human platelets induced by Ca^{+2}, *Eur. J. Biochem.*, 143, 437, 1984.

41. **Haslam, R. J. and Davidson, M. M. L.**, Potentiation by thrombin of the secretion of serotonin from permeabilized platelets equilibrated with Ca^{2+} buffers, *Biochem. J.*, 222, 351, 1984.

42. **Ashby, B., Kowalska, M. A., Wernick, E., Rigmaiden, M., Daniel, J. L., and Smith, J. B.**, Differences in the mode of action of 1-oleoyl-2-acetylglycerol and phorbol ester in platelet activation, *J. Cyclic Nucleotide Protein Phosphorylation Res.*, 10, 473, 1985.

43. **Lapetina, E. G., Reep, B., Ganong, B. R., and Bell, R. M.**, Exogenous sn-1,2-diaclyglycerols containing saturated fatty acids function as bioregulators of protein kinase C in human platelets, *J. Biol. Chem.*, 260, 1358, 1985.

44. **Johnson, P. C., Ware, J. A., Cliveden, P. B., Smith, M., Dvorak, A. M., and Salzman, E. W.**, Measurement of ionized calcium in blood platelets with the photoprotein aequorin. Comparison with quin 2, *J. Biol. Chem.*, 260, 2069, 1985.

45. **Hatayama, K., Kambayashi, J., Nakamura, K., Onshiro, T., and Mori, T.**, Fluorescent Ca^{2+}-indicator quin 2 as an intracellular Ca^{2+} antagonist in platelet reaction, *Thromb. Res.*, 38, 505, 1985.

46. **Hart, G. R., Ray, K. P., and Wallis, M.**, Use of quin 2 to measure calcium concentrations in ovine anterior pituitary cells and the effect of quin 2 on secretion of growth hormone and prolactin, *FEBS Lett.*, 203, 77, 1986.

47. **Lew, P. D., Wollheim, C. B., Waldvogel, F. A., and Pozzan, T.**, Modulation of cytosolic-free calcium transients by changes in intracellular calcium buffering capacity: correlation with exocytosis and O_2^--production in human neutrophils, *J. Cell Biol.*, 99, 1212, 1984.

48. **Nishizuka, Y.**, The role of protein kinase C in cell surface signal transduction and tumour promotion, *Nature (London)*, 308, 693, 1984.

49. **Sekar, M. C. and Hokin, L. E.**, The role of phosphoinositides in signal transduction, *J. Membrane Biol.*, 89, 193, 1986.

50. **Hokin, L. E.**, Receptors and phosphoinositide-generated second messengers, *Ann. Rev. Biochem.*, 54, 205, 1985.

51. **Rink, T. J., Sanchez, A., and Hallam, T. J.**, Diacylglycerol and phorbol ester stimulate secretion without raising cytoplasmic free calcium in human platelets, *Nature (London)*, 305, 317, 1983.

52. **Watson, S. P., Reep, B., McConnell, T. R., and Lapetina, E. G.**, Collagen stimulates [³H]inositol trisphosphate formation in indomethacin-treated human platelets, *Biochem. J.*, 226, 831, 1985.

53. **Krishnamurthi, S., Joseph, S. K., and Kakkar, V. V.**, Lack of inhibition of thrombin-induced rise in intracellular Ca^{2+} levels and 5-hydroxytryptamine secretion by 1-oleoyl-2-acetylglycerol in human platelets, *FEBS Lett.*, 196, 365, 1986.

54. **Ware, J. A., Johnson, P. C., Smith, M., and Salzman, E. W.**, Aequorin detects increased cytoplasmic calcium in platelets stimulated with phorbol ester or diacylglycerol, *Thromb. Res.*, 133, 98, 1985.

55. **Sha'afi, R. I., White, J. R., Molski, T. F. P., Shefcyk, J., Volpi, M., Naccache, P. H., and Feinstein, M. B.**, Phorbol-12-myristate 13-acetate activates rabbit neutrophils without an apparent rise in the level of intracellular free calcium, *Biochem. Biophys. Res. Commun.*, 114, 638, 1983.

56. **Barrowman, M. M., Cockcroft, S., and Gomperts, B. D.**, Two roles for guanine nucleotides in the stimulus-secretion sequence of neutrophils, *Nature (London)*, 319, 504, 1986.

57. **Roos, A., and Boron, W. F.**, Intracellular pH, *Physiol. Rev.*, 61, 296, 1981.

58. **Mahnensmith, R. L. and Aronson, P. S.**, The plasma membrane sodium-hydrogen exchanger and its role in physiological and pathophysiological processes, *Circ. Res.*, 56, 773, 1985.

59. **Aronson, P. S.**, Kinetic properties of the plasma membrane Na^+/H^+ exchanger, *Annu. Rev. Physiol.*, 47, 545, 1985.

60. **Cassel, D., Katz, M., and Rotman, M.**, Depletion of cellular ATP inhibits Na^+/H^+ antiport in cultured human cells, *J. Biol. Chem.*, 261, 5460, 1986.

61. **Johnson, R. G., Carty, S. E., Fingerhood, B. J., and Scarpa, A.**, The internal pH of mast cell granules, *FEBS Lett.*, 120, 75, 1980.

62. **Wilkins, J. A. and Salganicoff, L.**, Participation of a transmembrane proton gradient in 5-hydroxytryptamine transport by platelet dense granules and dense-granule ghosts, *Biochem. J.*, 1981, 113, 1981.

63. **Grinstein, S. and Furuya, W.**, A Mg^{2+} stimulated ATPase in platelet α-granule membranes: possible involvement in proton translocation, *Arch. Biochem. Biophys*, 218, 502, 1982.

64. **Carty, S. E., Johnson, R. G., and Scarpa, A.**, Electrochemical proton gradient in dense granules isolated from anterior pituitary, *J. Biol. Chem.*, 257, 7269, 1982.

65. **Johnson, R. G., Carty, S., and Scarpa, A.,** A model of biogenic amine accumulation into chromaffin granules and ghosts based on coupling to the electrochemical proton gradient, *Fed. Proc.,* 41, 2746, 1982.

66. **Knoth, J., Zallakian, M., and Njus, D.,** Mechanism of proton-linked monoamine transport in chromaffin granule ghosts, *Fed. Proc.,* 41, 2742, 1982.

67. **Busa, W. B. and Nuccitelli, R.,** Metabolic regulation via intracellular pH, *Am. J. Physiol.,* 246, R409, 1984.

68. **Moolenaar, W. H.,** Effects of growth factors on intracellular pH regulation, *Annu. Rev. Physiol.,* 48, 363, 1986.

69. **Grinstein, S. and Rothstein, A.,** Mechanism of regulation of the Na^+/H^+ exchanger, *J. Membrane Biol.,* 90, 1, 1986.

70. **Grinstein, S. and Furuya, W.,** Amiloride-sensitive Na^+/H^+ exchange in human neutrophils: mechanism of activation by chemotactic factors, *Biochem. Biophys. Res. Commun.,* 122, 755, 1984.

71. **Grinstein, S., Cohen, S., Goetz, J. D., Rothstein, A., and Gelfand, E. W.,** Characterization of the activation of Na^+/H^+ exchange in lymphocytes by phorbol esters: change in cytoplasmic pH dependence of the antiport, *Proc. Natl. Acad. Sci. U.S.A.,* 82, 1429, 1985.

72. **Grinstein, S., Cohen, S., Goetz, J. D., and Rothstein, A.,** Osmotic and phorbol ester-induced activation of Na^+/H^+ exchange: possible role of protein phosphorylation in lymphocyte volume regulation, *J. Cell Biol.,* 101, 269, 1985.

73. **Simchowitz, L.,** Chemotactic factor-induced activation of Na^+/H^+ exchange in human neutrophils. I. Sodium fluxes, *J. Biol. Chem.,* 260, 13237, 1985.

74. **Simchowitz, L.,** Chemotactic factor-induced activation of Na^+/H^+ exchange in human neutrophils. II. Intracellular pH changes, *J. Biol. Chem.,* 260, 13248, 1985.

75. **Horne, W. C., Norman, N. E., Schwartz, D. B., and Simons, E. R.,** Changes in cytoplasmic pH and in membrane potential in thrombin-stimulated human platelets, *Eur. J. Biochem.,* 120, 295, 1981.

76. **Siffert, W., Fox, G., Mückenhoff, K., and Scheid, P,** Thrombin stimulates Na^+/H^+ exchange across the human platelet plasma membrane, *FEBS Lett.,* 172, 272, 1984.

77. **Dufresne, M., Bastie, M.-J., Vaysse, N., Creach, Y, Hollande, E., and Ribet, A.,** The amiloride sensitive Na^+/H^+ antiport in guinea pig pancreatic acini. Characterization and stimulation by caerulein, *FEBS Lett.,* 187, 126, 1985.

78. **Horne, W. C. and Simons, E. R.,** Effects of amiloride on the response of human platelets to bovine α-thrombin, *Thromb. Res.,* 13, 599, 1978.

79. **Connolly, T. M. and Limbird, L. E.,** Removal of extraplatelet Na^+ eliminates indomethacin-sensitive secretion from human platelets stimulated by epinephrine, ADP, and thrombin, *Proc. Natl. Acad. Sci. U.S.A.,* 80, 5320, 1983.

80. **Connolly, T. M. and Limbird, L. E.,** The influence of Na^+ on the α_2-adrenergic receptor system of human platelets, *J. Biol. Chem.,* 258, 3907, 1983.

81. **Sweatt, J. D., Johnson, S. L., Cragoe, E. J., and Limbird, L. E.,** Inhibitors of Na^+/H^+ exchange block stimulus-provoked arachidonic acid release in human platelets, *J. Biol. Chem.,* 260, 12910, 1985.

82. **Sweatt, J. D., Connolly, T. M., Cragoe, E. J., and Limbird, L. E.,** Evidence that Na^+/H^+ exchange regulates receptor-mediated phospholipase A_2 activation in human platelets, *J. Biol. Chem.,* 261, 8667, 1986.

83. **Kannagi, R. and Koizumi, K.,** Phospholipid-deacylating enzymes of rabbit platelets, *Arch. Biochem. Biophys.,* 196, 534, 1979.

84. **Siffert, W. and Akkerman, J. W. N.,** Activation of sodium-proton exchange is a prerequisite for Ca^{+2} mobilization in human platelets, *Nature (London),* 325, 456, 1987.

85. **Leven, R. M., Gonnella, P. A., Reeber, N. G., and Nachmias, V. T.,** Platelet shape change and cytoskeletal assembly: effect of pH and monovalent cation inophores, *Thromb. Haemostas.,* 49, 320, 1983.

86. **Brass, L. F. and Joseph, S. K.,** A role for inositol trisphosphate in intracellular Ca^{2+} mobilization and granule secretion in platelets, *J. Biol. Chem.,* 260, 15172, 1986.

87. **Feinstein, M. B., Egan, J. J., Sha'afi, R. I., and White, J.,** The cytoplasmic concentration of free calcium in platelets is controlled by stimulators of cyclic AMP production (PGD_2, PGE_1, Forskolin), *Biochem. Biophys. Res. Commun.,* 113, 598, 1983.

88. **Zavoico, G. B. and Feinstein, M. B.,** Cytoplasmic Ca^{2+} in platelets is controlled by cyclic AMP: antagonism between stimulators and inhibitors of adenylate cyclase, *Biochem. Biophys. Res. Commun.,* 120, 579, 1984.

89. **Käser-Glanzmann, R., Jakobova, M., George, J. N., and Lüscher, E. F.,** Stimulation of calcium uptake in platelet membrane vesicles by adenosine 3,5-cyclic monophosphate and protein kinase, *Biochim. Biophys. Acta,* 466, 429, 1977.

90. **Käser-Glanzmann, R., Gerber, E., and Lüscher, E. F.,** Regulation of the intracellular calcium level in human blood platelets; cyclic adenosine 3,5-monophosphate dependent phosphorylation of a 22.000 dalton component in isolated Ca^{2+}-accumulating vesicles, *Biochim. Biophys. Acta,* 558, 344, 1979.

91. **Watson, S. P., McConnell, R. T., and Lapetina, E. G.,** The rapid formation of inositol phosphates in human platelets by thrombin is inhibited by prostacyclin, *J. Biol. Chem.,* 259, 13199, 1984.
92. **Della Bianca, V., De Togui, P., Grezeskowiak, M., Vincentini, L. M., and Di Virgilio, F.,** Cyclic AMP inhibition of phosphoinositide turnover in human neutrophils, *Biochim. Biophys. Acta,* 886, 441, 1986.
93. **Haslam, R. J., Sulama, S. E., Fox, J. E. B., Lynham, J. A., and Davidson, M. M. L.,** Roles of cyclic nucleotides and of protein phosphorylation in the regulation of platelet function, in *Platelets: Cellular Response Mechanisms and Their Biological Significance,* Rotman, A., Meyer, F. E., Gitler, C., and Silberberg, A., Eds., John Wiley & Sons, New York, 1980, 218.
94. **MacIntyre, D. E., Bushfield, M., and Shaw, A. M.,** Regulation of platelet cytosolic free calcium by cyclic nucleotides and protein kinase C, *FEBS Lett.,* 188, 383, 1985.
95. **Waldmann, R., Bauer, S., Gobel, C., Hofmann, F., Jakobs, K. H., and Walter, U.,** Demonstration of cGMP-dependent protein kinase and cGMP-dependent phosphorylation in cell-free extracts of platelets, *Eur. J. Biochem.,* 158, 203, 1986.
96. **Thibault, G., Garcia, R., Gutkowska, J., Genest, J., and Cantin, M.,** Atrial natriuretic factor. A newly discovered hormone with significant clinical implications, *Drugs,* 31, 369, 1986.
97. **Palluk, R., Gaida, W., and Hoefke, W.,** Atrial natriuretic factor, *Life Sci.,* 36, 1415, 1985.
98. **Meisheri, K. D., Taylor, C. J., and Saneii, H.,** Synthetic atrial peptide inhibits intracellular calcium release in smooth muscle, *Am. J. Physiol.,* 250, C171, 1986.
99. **Ignarro, L. J., Wood, K. S., Harbison, R. G., and Kadowitz, P. J.,** Atriopeptin II relaxes and elevates cGMP in bovine pulmonary artery but not vein, *J. Appl. Physiol.,* 60, 1128, 1986.
100. **Rapoport, R. M., Ginsburg, R., Waldman, S. A., and Murad, F.,** Effects of atriopeptins on relaxation and cyclic GMP levels in human coronary artery in vitro, *Eur. J. Pharmacol.,* 124, 193, 1986.
101. **Silver, P. J., Kocmund, S. M., and Pinto, P. B.,** Enhanced phosphorylation of arterial particulate proteins by cyclic nucleotides and human atrial natriuretic factor, *Eur. J. Pharmacol.,* 122, 385, 1986.
102. **Gutkowska, J., Horky, K., Schiffrin, E. L., Thibault, G., Garcia, R., de Lean, A., Hamet, P., Tremblay, J., Anannd-Stivastava, M. B., Januszewicz, P., Genest, J., and Cantin, M.,** Atrial natriuretic factor: radioimmunoassay and effects on adrenal and pituitary glands, *Fed. Proc.,* 45, 2101, 1986.

Chapter 10

THE ROLE OF PROTEIN KINASE C IN SECRETORY RESPONSES

Katsuhiko Ase, Hideaki Nomura, Masayoshi Go, Tadashi Hashimoto, Kouji Ogita, Ushio Kikkawa, and Yasutomi Nishizuka

TABLE OF CONTENTS

I. INTRODUCTION

Response of inositol phospholipids to the activation of cell surface receptors was first recognized by Hokin and Hokin[1] who showed that acetylcholine induces rapid breakdown and resynthesis of phosphatidylinositol (PI) in some secretory tissues such as pancreas. It later became evident that such enhanced turnover of PI occurs in many cell types in response to a wide variety of external signals.[2-5] Inositol phospholipids contain most frequently a 1-stearoyl-2-arachidonyglycerol backbone, with a small portion of the inositols containing one phosphate at position 4 (phosphatidylinositol 4-phosphate, PIP) or two phosphates at positions 4 and 5 (phosphatidylinositol 4,5-bisphosphate, PIP_2). These minor phospholipids each represent 1 to 2% of the total inositol phospholipids and are produced through sequential phosphorylation of PI. In earlier studies PI was regarded as the primary target for signal-induced breakdown, but PIP_2 has been more intensely studied in recent years because it generally disappears more rapidly than PI in stimulated cells, and its water-soluble product, inositol-1,4,5-trisphosphate (I-1,4,5-P_3), has been shown to serve as an intracellular mediator for the mobilization of Ca^{2+} from internal stores.[6] A series of studies in our laboratory has provided evidence that 1,2-diacylglycerol, the other product of PIP_2 breakdown, remains in the membrane and initiates the activation of a specialized protein kinase, protein kinase C.[7,8] Thus, the information from extracellular signals is transduced across the membrane via a bifurcating pathway, Ca^{2+} mobilization on one hand and protein phosphorylation on the other as outlined in Figure 1. The importance of protein phosphorylation in this signal pathway was first demonstrated for serotonin release reaction from platelets, and subsequently for physiological cellular responses in a wide variety of tissues. This chapter will briefly summarize possible roles of protein kinase C in secretory responses. Several other aspects of this protein kinase C have been reviewed.[7-12]

II. SOME PROPERTIES OF PROTEIN KINASE C

Protein kinase C is distributed in many tissues and organs, with platelets and brain having the highest activity.[13,14] In brain tissues a large quantity of the enzyme is associated with synaptic membranes, whereas in most other tissues, including platelets, the enzyme is present apparently in the soluble fraction as an inactive form.[15] Recent immunocytochemical analysis with monoclonal antibodies against protein kinase C suggests that this enzyme is absent or poorly represented in the nucleus.[16] Protein kinase C preparations obtained from various tissues are apparently similar to one another in their kinetic and catalytic properties. However, the enzyme isolated from rat brain soluble fraction usually reveals a doublet when subjected to sodium dodecyl sulfate (SDS)-polyacrylamide gel electrophoresis.[17] It is likely that more than one gene may exist for this protein kinase.

The molecular weight of protein kinase C is about 82,000 upon SDS-polyacrylamide gel electrophoresis. The Stokes radius is 42 Å, and the molecular weight is 87,000 as estimated by gel filtration analysis. The sedimentation coefficient is 5.1 S which corresponds to a molecular weight of 77,000. The enzyme is composed of a single polypeptide chain with no subunit structure. The frictional ratio of the enzyme is calculated to be 1.6, indicating an asymmetric nature of the molecule. The isoelectric point of the enzyme is pH 5.6. The optimum pH range for the activity is 7.5 to 8.0 with Tris/acetate as a test buffer. Mg^{2+} is essential for the catalytic activity with an optimum range of about 5 to 10 mM. The K_m value for ATP is about 6×10^{-6} M. GTP cannot serve as a phosphate donor. Like many other protein kinases the enzyme phosphorylates itself, but the significance of this autophosphorylation is not known.

Protein kinase C requires both Ca^{2+} and phospholipid for its activation. When diacylglycerol is produced in membranes, it dramatically increases the affinity of the enzyme for Ca^{2+}, thereby rendering it fully active without a net increase in the Ca^{2+} concentration.[18,19]

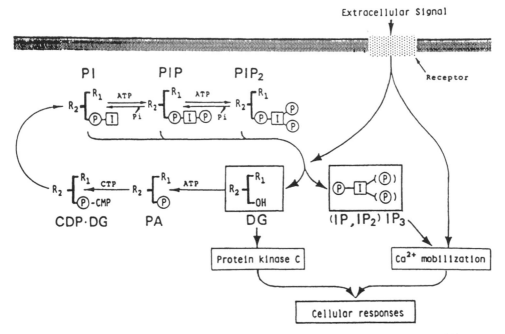

FIGURE 1. Inositol phospholipid turnover and cell surface signal transduction. PI, phosphatidylinositol; PIP, phosphatidylinositol 4-phosphate; PIP_2, phosphatidylinositol 4,5-bisphosphate; DG, 1,2-diacylglycerol; IP_3, inositol-1,4,5-trisphosphate; PA, phosphatidic acid; CDP-DG, cytidine diphosphodiacylglycerol; I, inositol moiety; P, phorsphoryl group.[7]

Thus, the activation of this protein kinase is biochemically dependent on, but physiologically independent of a change in Ca^{2+} concentration. Among various phospholipids tested only phosphatidylserine is indispensable. Other phospholipids such as phosphatidylethanolamine, phosphatidylinositol, phosphatidylcholine, and sphingomyelin are all inactive. However, these inactive phospholipids show positive or negative cooperativity in the activation of protein kinase C when they are mixed with phosphatidylserine. For instance, phosphatidylethanolamine increases further the affinity of protein kinase C for Ca^{2+}, making it fully active at 10^{-7} M Ca^{2+}. Both phosphatidylcholine and sphingomyelin, on the other hand, decrease the affinity of protein kinase C for Ca^{2+}.[19] Presumably, the asymmetric distribution of various phospholipids in the membrane phospholipid bilayer, wherein phosphatidylserine is enriched in the inner leaflet, may take part in the activation of the enzyme.

Various diacylglycerols are able to activate protein kinase C, and no absolute specificity is observed for the 1-stearoyl-2-arachidonyl structure. Diacylglycerols most active in vitro contain at least one unsaturated fatty acyl moiety at either position 1 or 2.[18] Some diacylglycerols having relatively short carbon chains such as dioctanoylglycerol, didecanoylglycerol,[20] and 1-palmitoyl-2-acetylglycerol[21] are also active in supporting enzyme activation. It is worth noting that only 1,2-*sn*- but not 2,3-*sn*-diacylglycerols are active, suggesting that a highly specific lipid-protein interaction is needed for this enzyme activation.[22] The detailed biochemical mechanism of this enzyme activation is largely unknown.

Protein kinase C can also be activated by proteolysis with Ca^{2+}-dependent protease (calpain) or trypsin.[23,24] Limited proteolysis with calpain results in the formation of M_r 51,000 component that carries the enzymatic activity. The catalytic activity of this fragment is totally independent of Ca^{2+}, phospholipid, and diacylglycerol. When protein kinase C is attached to the membrane, it is more susceptible to this limited proteolysis.[24] It has been proposed that such proteolytic activation of protein kinase C may occur in intact platelets[25] and neutrophils.[26] The true physiological significance of this proteolysis has not yet been established.

III. ACTIVATION OF PROTEIN KINASE C BY RECEPTOR BYPASS

To explore the role of protein kinase C in stimulus-response coupling, a synthetic dia-cylglycerol, 1-oleoyl-2-acetylglycerol, has been used, since it is readily intercalated into intact cell membranes, dispersed in the phospholipid bilayer, and activates protein kinase C directly.[27] Later, 1,2-dioctanoylglycerol and 1,2-didecanoylglycerol are also shown to be effective permeable diacylglycerols.[20] Synthetic diacylglycerols having 2,3-*sn*-configuration are not active for intact cell systems.[28] The diacylglycerols derived from triacylglycerol by the action of lipoprotein lipase[29] and a heparin-releasable hepatic lipase[30] have been proposed to show 2,3-*sn*-configuration. When added to intact cells, the active permeable diacylgly-cerols mentioned above are metabolized rapidly and converted to the corresponding phos-phatidic acids and probably further to inositol phospholipids.[27,31]

Tumor-promoting phorbol esters, such as 12-*O*-tetradecanoylphorbol-13-acetate (TPA), have a structure in their molecules similar to diacylglycerol and activate protein kinase C both in vitro and in vivo.[32,33] Kinetic analysis indicates that TPA, like diacylglycerol, dramatically increases the affinity of this enzyme for Ca^{2+}, resulting in its full activation at physiological concentrations of this divalent cation. Several lines of evidence provided by many laboratories seem to suggest that protein kinase C is probably the prime target of tumor promoters.[7,11]

IV. SYNERGISM OF CALCIUM ION AND PROTEIN KINASE C

Ca^{2+} has been long known to be required for secretory responses. Intracellular Ca^{2+} concentration is normally sustained within the range of 0.1 to 0.4 μM. The plasma membrane is relatively impermeable to Ca^{2+}, and Ca^{2+} that enters the cell is pumped back out by both a Ca^{2+}-transport ATPase and an Na^+/Ca^{2+}-exchange system to maintain cellular Ca^{2+} homeostasis.[34] I-1,4,5-P_3 appears to act on internal Ca^{2+} stores, probably in a compartment of the endoplasmic reticulum through interaction with its receptor.[35,36] The inositol tris-phosphate, once produced, disappears very rapidly, and a major mechanism for terminating this signal pathway is thought to be removal of the 5-phosphate by the action of a specific phosphatase. Although there might be more than one mechanism eventually leading to the required intracellular Ca^{2+} concentration, the Ca^{2+} signal in most tissues is transient and returns quickly to basal or even below basal levels. In experiments with the photoprotein aequorin this Ca^{2+} spike has been successfully observed.[37-39]

Similarly, the formation of diacylglycerol in membranes is also transient, and this neutral lipid disappears within a few seconds or, at most, a minute. This rapid disappearance is due both to its conversion back to inositol phospholipids by way of phosphatidic acid (PI turnover) and to its further degradation to provide arachidonic acid for generating another class of messengers such as prostaglandins. Thus, protein kinase C reveals its activity only for a short time after the stimulation of the receptor, and is inactivated rapidly with the disap-pearance of diacylglycerol and presumably also with the proteolytic degradation of the protein kinase molecule.

Under appropriate conditions, Ca^{2+} mobilization and protein kinase C activation can be induced selectively and independently by the application of a Ca^{2+}-ionophore such as A23187 or ionomycin for the former and a permeable diacylglycerol or phorbol ester for the latter. Using this procedure it has been repeatedly shown that Ca^{2+} mobilization and protein kinase C activation are both essential to elicit full cellular responses. Such a role of protein kinase C in stimulus-response coupling was first demonstrated for the release of serotonin from platelets.[27] Figure 2 shows the the synergistic effects of A23187 and permeable diacylglycerol on platelet release reactions. In the presence of appropriate concentrations of this Ca^{2+} ionophore, 1-oleoyl-2-acetylglycerol induced full release reaction of serotonin as well as

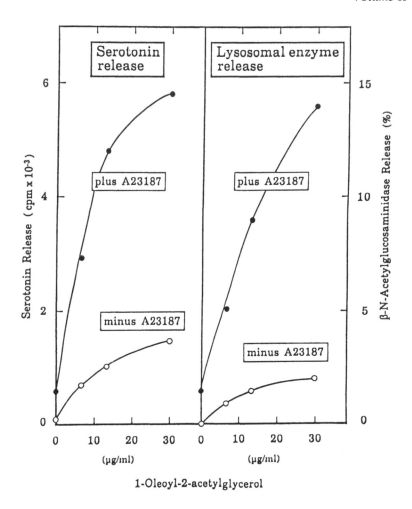

FIGURE 2. Synergistic effect of permeable diacylglycerol and Ca^{2+} ionophore on platelet release reactions. Human platelets were labeled with [^{14}C] serotonin and then stimulated in the presence or absence of A23187 (0.4 μM) as indicated. The amount of serotonin release was estimated from its radioactivity. β-N-Acetylglucosaminidase was assayed colorimetrically with p-nitrophenyl-N-acetyl-β-glucosaminide.[40]

lysosomal enzymes. Under these conditions the Ca^{2+} ionophore itself showed little effect on the release reactions. Figure 3 shows the stereospecificity of diacylglycerol for the activation of purified enzyme and intact cell systems. In the presence of Ca^{2+}-ionophore, 1,2-sn-dioctanoylglycerol caused full activation of the release of serotonin. 2,3-sn-Dioctanoylglycerol and 1,3-dioctanoylglycerol were inactive. Thus, the diacylglycerol which functions in the stimulus-response coupling possesses a 1,2-sn-glycerol backbone, and other isomers are not involved in the signal transduction through the protein kinase C pathway. The role of protein kinase C in release, secretion, and exocytosis of cellular constituents has been shown for a variety of blood cells and endocrine, exocrine, and nervous tissues as shown in Table 1. However, the biochemical basis of such actions of protein kinase C in each tissue is not fully substantiated.

V. FEEDBACK CONTROL AND RELATION TO OTHER SIGNALING SYSTEMS

In biological systems positive signals are usually followed by immediate feedback control

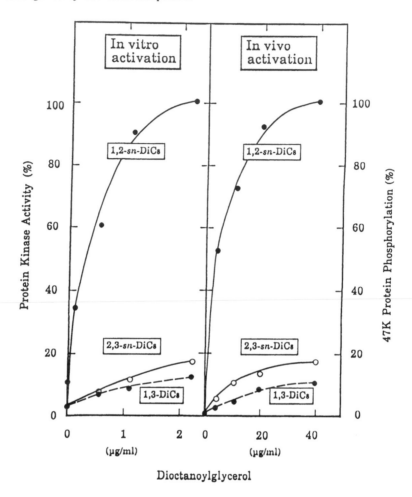

FIGURE 3.   Stereospecificity of diacylglycerol for activation of protein kinase C. In in vitro reactions homogeneous protein kinase C purified from rat brain was assayed in the presence of phosphatidylserine, Ca^{2+}, and diacylglycerol. In in vivo reactions, human platelets were isolated and labeled with ^{32}Pi and then stimulated by diacylglycerol as indicated. The phosphorylation of 47K protein, a specific platelet protein kinase C substrate, was determined by measuring the incorporation of radioactive phosphate into the protein, which was separated by sodium dodecyl sulfate-polyacrylamide gel electrophoresis and subjected to autoradiography, followed by densitometric tracing at 430 nm. Detailed conditions will be described elsewhere. 1,2-sn-DiC$_8$, 1,2-sn-dioctanoylglycerol; 2,3-sn-DiC$_8$, 2,3-sn-dioctanoylglycerol; and 1,3-DiC$_8$, 1,3-dioctanoylglycerol.

to prevent over-response. A major function of protein kinase C may be related to such feedback control, termed down-regulation. Recent evidence obtained for platelets,[88-91] strongly suggests that protein kinase C exerts negative feedback control on the thrombin-receptor, and inhibits the release reaction. Similar feedback control by protein kinase C over its own receptor that is related to inositol phospholipid breakdown has been suggested for various cell types such as neutrophils,[92] basophils,[47] and PC12 cells.[93] The biochemical basis of such a negative feedback control remains to be clarified.

Feedback inhibition of Ca^{2+} mobilization, that is related to PI turnover, is also reported.[94-99] Protein kinase C may play a role in extrusion of Ca^{2+} immediately after its mobilization into the cytosol, and the Ca^{2+}-transport ATPase is a possible target of protein kinase

Table 1
POTENTIAL ROLES OF PROTEIN KINASE C IN
SECRETORY RESPONSES

Tissues and cells	Secretory responses	Ref.
Blood cell systems		
Platelets	Serotonin	27, 41, 42
	Lysosomal enzyme	40, 42
	Arachidonate	43
Neutrophils	Lysosomal enzyme	40, 44—46
Basophils	Histamine	47
Mast cells	Histamine	48, 49
Endocrine systems		
Adrenal medulla	Catecholamine	50—53
	Arachidonate	52
Adrenal cortex	Aldosterone	54
Pancreatic islets	Insulin	55—59
	Glucagon	60
Insulinoma cells	Insulin	61
Pituitary cells	Anterior pituitary hormone	62
	Growth hormone	63—65
	Luteinizing hormone	64—68
	Prolactin	69, 70
	Adrenocorticotropin	71
	Thyrotropin	72
Parathyroid cells	Parathyroid hormone	73—75
Thyroid C cells	Calcitonin	76
Exocrine systems		
Pancreas	Amylase	77, 78
Parotid gland	Protein	79
Gastric gland	Pepsinogen	80
Parietal cells	Acid	81
Alveolar cells	Surfactant	82
Nervous systems		
Ileal nerve endings	Acetylcholine	83
Neuromuscular preparation	Transmitter	84
Caudate nucleus	Acetylcholine	85
PC12 cells	Dopamine	86
Brain neurons	Dopamine	87

C. Studies with some cell types such as platelets[38,39] have shown that, when receptors are stimulated, the appearance of Ca^{2+} is transient as noted above. In various cell types, the cytosolic Ca^{2+} concentration is decreased by the addition of phorbol ester.[47,99-103] With neutrophils it has been proposed that Ca^{2+}-transport ATPase is activated by the stimulation of protein kinase C.[103] However, more information is needed to clarify the role, if any, of this protein kinase in maintaining Ca^{2+}-homeostasis under physiological processes.

Another major signaling pathway for the control of cellular functions is mediated through cAMP. Most tissues seem to have two major receptors for transducing information across the membrane, one is related to cAMP, while the other induces turnover of inositol phospholipids as well as mobilization of Ca^{2+}. Stimulation of the latter class of receptors normally leads to the release of arachidonate, and often increases cGMP. Thus, protein kinase C activation, Ca^{2+} mobilization, arachidonate release, and cGMP formation appear to be integrated into a single receptor cascade.

Cellular responses may be divided into several modes as given schematically in Figure 4. In *bidirectional control systems* the two classes of receptors appear to counteract each other. In some cells such as platelets and neutrophils, the signals that induce inositol phos-

Bidirectional control systems

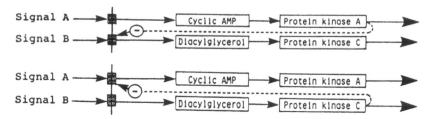

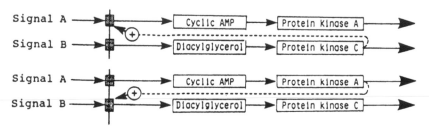

Monodirectional control systems

FIGURE 4.   Interaction of two major signaling systems.[11]

pholipid breakdown promote the activation of cellular functions, but the signals that produce cAMP usually antagonize such activation. In those cells the signal-induced breakdown of inositol phospholipids and the subsequent events eventually leading to cellular responses are all profoundly blocked by cAMP.[7,8] Inversely, in another group of cell types such as erythrocytes[104,105] and Leydig cells,[106-108] protein kinase C inhibits and desensitizes the adenylate cyclase system. In *monodirectional control systems* one class of receptors may potentiate the other. In several other cell types including pinealocytes[109] and pituitary cells,[110,111] protein kinase C greatly potentiates cAMP production. There is no obvious example as yet of a tissue in which cAMP potentiates signal-induced turnover of inositol phospholipids. However, these two signal transduction pathways probably act in concert in some endocrine tissues such as pancreatic islets.[55] The evidence presented thus far is still incomplete, but it is reasonable to assume that various combinations of the two receptor systems may operate positively and sometimes be intensified in many physiological processes.

VI. CONCLUSION

The present article briefly summarizes some aspects of protein kinase C. The enzyme appears to play crucial roles in stimulus-response coupling in many cells and tissues. Perhaps, the signal-induced breakdown of inositol phospholipids initiates a cascade of events, starting with Ca^{2+} mobilization and protein kinase C activation. Each of the signal pathways may play diverse roles in controlling biochemical reactions. The protein phosphorylation catalyzed by this protein kinase may exert profound modulation of various Ca^{2+}-mediated processes, particularly secretion and exocytosis. Further exploration of the target proteins of this protein kinase in individual cell types is of great importance for understanding the molecular basis of transmembrane control of secretory responses.

ACKNOWLEDGMENTS

This research has been supported by the Ministry of Education, Science and Culture,

Japan; Muscular Dystrophy Association; Yamanouchi Foundation for Research on Metabolic Disorders; Merck Sharp & Dohme Research Laboratories; Central Research Laboratories, Ajinomoto Company; Biotechnology Laboratories, Takeda Chemical Industries; and Meiji Institute of Health Sciences.

REFERENCES

1. **Hokin, M. R. and Hokin, L. E.**, Enzyme secretion and the incorporation of ^{32}P into phospholipids of pancreatic slices, *J. Biol. Chem.*, 203, 967, 1953.
2. **Hokin, M. R. and Hokin, L. E.**, Interconversion of phosphatidylinositol and phosphatidic acid involved in the response to acetylcholine in the salt gland, in *The Metabolism and Physiological Significance of Lipids*, Dawson, R. M. C. and Rhodes, D. N., Eds., John Wiley & Sons, New York, 1964, 423.
3. **Michell, R. H.**, Inositol phospholipids and cell surface receptor function, *Biochim. Biophys. Acta*, 415, 81, 1975.
4. **Hawthorne, J. N. and Pickard, M. R.**, Phospholipid in synaptic function, *J. Neurochem.*, 32, 5, 1979.
5. **Irvine, R. F., Dawson, R. M. C., and Freinkel, N.**, Stimulated phosphatidylinositol turnover: A brief proposal, in *Contemporary Metabolism*, Vol. 2, Freinkel, N., Ed., Plenum Press, New York, 1982, 301.
6. **Berridge, M. J. and Irvine, R. F.**, Inositol trisphosphate, a novel second messenger in cellular signal transduction, *Nature (London)*, 312, 315, 1984.
7. **Nishizuka, Y.**, The role of protein kinase C in cell surface signal transduction and tumour promotion, *Nature (London)*, 308, 693, 1984.
8. **Nishizuka, Y.**, Turnover of inositol phospholipids and signal transduction, *Science*, 225, 1365, 1984.
9. **Rasmussen, H. and Barrett, P. Q.**, Calcium messenger system: an integrated view, *Physiol. Rev.*, 64, 938, 1984.
10. **Williamson, J. R., Cooper, R. H., Joseph, S. K., and Thomas, A. P.**, Inositol trisphosphate and diacylglycerol and intracellular second messengers in liver, *Am. J. Physiol.*, 248, C203, 1985.
11. **Nishizuka, Y.**, Perspectives on the role of protein kinase C in stimulus-response coupling, *J. Natl. Cancer Inst.*, 76, 363, 1986.
12. **Kikkawa, U. and Nishizuka, Y.**, Protein kinase C, in *The Enzymes*, Vol. 17., Krebs, E. G. and Boyer, P. D., Eds., Academic Press, Orlando, Fla., 1986, 167.
13. **Kuo, J. F., Andersson, R. G. G., Wise, B. C., Mackerlova, L., Solomonsson, I., Brackett, N. L., Katoh. N., Shoji, M., and Wrenn, W. R.**, Calcium-dependent protein kinase: widespread occurrence in various tissues and phyla of the animal kingdom and comparison of effects of phospholipid, calmodulin, and trifluoperazine, *Proc. Natl. Acad. Sci. U.S.A.*, 77, 7039, 1980.
14. **Minakuchi, R., Takai, Y., Yu, B., and Nishizuka, Y.**, Widespread occurrence of calcium-activated, phospholipid-dependent protein kinase in mammalian tissues, *J. Biochem.*, 89, 1651, 1981.
15. **Kikkawa, U., Takai, Y., Minakuchi, R., Inohara, S., and Nishizuka, Y.**, Calcium-activated, phospholipid-dependent protein kinase from rat brain. Subcellular distribution, purification, and properties, *J. Biol. Chem.*, 257, 13341, 1982.
16. **Kikkawa, U., Kitano, T., Saito, N., Fujiwara, H., Nakanishi, H., Kishimoto, A., Taniyama, K., Tanaka C., and Nishizuka, Y.**, Possible roles of protein kinase C in signal transduction in nervous tissues, *Prog. Brain Res.*, 69, 29, 1986.
17. **Kikkawa, U., Go, M., Koumoto, J., and Nishizuka, Y.**, Rapid purification of protein kinase C by high performance liquid chromatography, *Biochem. Biophys. Res. Commun.*, 135, 636, 1986.
18. **Kishimoto, A., Takai, Y., Mori, T., Kikkawa, U., and Nishizuka, Y.**, Activation of calcium and phospholipid-dependent protein kinase by diacylglycerol, its possible relation to phosphatidylinositol turnover, *J. Biol. Chem.*, 255, 2273, 1980.
19. **Kaibuchi, K., Takai, Y., and Nishizuka, Y.**, Cooperative roles of various membrane phospholipids in the activation of calcium-activated, phospholipid-dependent protein kinase, *J. Biol. Chem.*, 256, 7146, 1981.
20. **Lapetina, E. G., Reep, B., Ganong, B. R., and Bell, R. M.**, Exogenous *sn*-1,2-diacylglycerols containing saturated fatty acids function as bioregulators of protein kinase C in human platelets, *J. Biol. Chem.*, 260, 1358, 1985.
21. **Kajikawa, N., Kikkawa, U., Itoh, K., and Nishizuka, Y.**, Permeable diacylglycerol, its application to release reactions, and partial purification and properties of protein kinase C from platelets, *Methods Enzymol.*, in press.

22. **Rando, R. R. and Young, N.,** The stereospecific activation of protein kinase C, *Biochem. Biophys. Res. Commun.,* 122, 818, 1984.

23. **Inoue, M., Kishimoto, A., Takai, Y., and Nishizuka, Y.,** Studies on a cyclic nucleotide-independent protein kinase and its proenzyme in mammalian tissues. II. Proenzyme and its activation by calcium-dependent protease from rat brain, *J. Biol. Chem.,* 252, 7610, 1977.

24. **Kishimoto, A., Kajikawa, N., Shiota, M., and Nishizuka, Y.,** Proteolytic activation of calcium-activated, phospholipid-dependent protein kinase by calcium-dependent neutral protease, *J. Biol. Chem.,* 258, 1156, 1983.

25. **Tapley, P. M. and Murray, A. W.,** Evidence that treatment of platelets with phorbol ester causes proteolytic activation of Ca^{2+}-activated, phospholipid-dependent protein kinase, *Eur. J. Biochem.,* 151, 419, 1985.

26. **Melloni, E., Pontremoli, S., Michetti, M., Sacco, O., Sparatore, B., Salamino, F., and Horecker, B. L.,** Binding of protein kinase C to neutrophil membranes in the presence of Ca^{2+} and its activation by a Ca^{2+}-requiring proteinase, *Proc. Natl. Acad. Sci. U.S.A.,* 82, 6435, 1985.

27. **Kaibuchi, K., Takai, Y., Sawamura, M., Hoshijima, M., Fujikura, T., and Nishizuka Y.,** Synergistic functions of protein phosphorylation and calcium mobilization in platelet activation, *J. Biol. Chem.,* 258, 6701, 1983.

28. **Nomura, H., Nakanishi, H., Ase, K., Kikkawa, U., and Nishizuka, Y.,** Inositol phospholipid turnover in stimulus-response coupling, *Prog. Hemostasis Thromb.,* 8, 143, 1986.

29. **Morley, N. and Kuksis, A.,** Positional specificity of lipoprotein lipase, *J. Biol. Chem.,* 247, 6389, 1972.

30. **Åkesson, B., Gronowitz, S., and Hersloef, B.,** Stereospecificity of hepatic lipases, *FEBS Lett.,* 71, 241, 1976.

31. **Lapetina, E. G.,** Incorporation of synthetic 1,2-diacylglycerol into platelet phosphatidylinositol is increased by cyclic AMP, *FEBS Lett.,* 195, 111, 1986.

32. **Castagna, M., Takai, Y., Kaibuchi, K., Sano, K., Kikkawa, U., and Nishizuka, Y.,** Direct activation of calcium-activated, phospholipid-dependent protein kinase by tumor-promoting phorbol esters, *J. Biol. Chem.,* 257, 7847, 1982.

33. **Yamanishi, J., Takai, Y., Kaibuchi, K., Sano, K., Castagna, M., and Nishizuka, Y.,** Synergistic functions of phorbol ester and calcium in serotonin release from human platelets, *Biochem. Biophys. Res. Commun.,* 112, 778, 1983.

34. **Carafoli, E. and Zurini, M.,** The Ca^{2+}-pumping ATPase of plasma membranes: purification, reconstitution and properties, *Biochim. Biophys. Acta,* 683, 279, 1982.

35. **Spat, A., Fabiato, A., and Rubin, R. P.,** Binding of inositol trisphosphate by a liver microsomal fraction, *Biochem. J.,* 233, 929, 1986.

36. **Spat, A., Bradford, P. G., McKinney, J. S., Rubin, R. P., and Putney, Jr., J. W.,** A saturable receptor for ^{32}P-inositol-1,4,5-trisphosphate in hepatocytes and neutrophils, *Nature (London),* 319, 514, 1986.

37. **Morgan, J. P. and Morgan, K. G.,** Vascular smooth muscle: the first recorded Ca^{2+} transients, *Pflugers Arch.,* 395, 75, 1982.

38. **Johnson, P. C., Ware, J. A., Cliveden, P. B., Smith, M., Dvorak, A. M., and Salzman, E. W.,** Measurement of ionized calcium in blood platelets with the photoprotein aequorin, *J. Biol. Chem.,* 260, 2069, 1985.

39. **Kikkawa, U., Kitano, T., Saito, N., Kishimoto, A., Taniyama, K., Tanaka, C., and Nishizuka, Y.,** Role of protein kinase C in calcium-mediated signal transduction, in *Calcium and the Cell,* Evered, D. and Whelan, J., Ed., John Wiley & Sons, New York, 1986.

40. **Kajikawa, N., Kaibuchi, K., Matsubara, T., Kikkawa, U., Takai, Y., and Nishizuka, Y.,** A possible role of protein kinase C in signal-induced lysosomal enzyme release, *Biochem. Biophys. Res. Commun.,* 116, 743, 1983.

41. **Rink, T. J., Sanchez, A., and Hallam, T. J.,** Diacylglycerol and phorbol ester stimulate secretion without raising cytoplasmic free calcium in human platelets, *Nature (London),* 305, 317, 1983.

42. **Knight, D. E., Niggli, V., and Scurton, M. C.,** Thrombin and activators of protein kinase C modulate secretory responses of permeabilised human platelets induced by Ca^{2+}, *Eur. J. Biochem.,* 143, 437, 1984.

43. **Halenda, S. P., Zavoico, G. B., and Feinstein, M. B.,** Phorbol esters and oleoyl acetyl glycerol enhance release of arachidonic acid in platelets stimulated by Ca^{2+} ionophore A23187, *J. Biol. Chem.,* 260, 12484, 1985.

44. **White, J. R., Huang, C.-K., Hill, J. M. Jr., Naccache, P. H., Becker, E. L., and Sha'afi, R. I.,** Effect of phorbol 12-myristate 13-acetate and its analogue 4α-phorbol 12,13-didecanoate on protein phosphorylation and lysosomal enzyme release in rabbit neutrophil, *J. Biol. Chem.,* 259, 8605, 1984.

45. **O'Flaherty, J. T., Schmitt, J. D., McCall, C. E., and Wykle, R. L.,** Diacylglycerols enhance human neutrophil degranulation responses: relevance to a multiple mediator hypothesis of cell function, *Biochem. Biophys. Res. Commun.,* 123, 64, 1984.

46. **Hoult, J. R. and Nourshargh, S.,** Phorbol myristate acetate enhances human polymorphonuclear neutrophil release of granular enzymes but inhibits chemokinesis, *Br. J. Pharmacol.,* 86, 533, 1985.

47. **Sagi-Eisenberg, R., Lieman, H., and Pecht, I.,** Protein kinase C regulation of the receptor-coupled calcium signal in histamine-secreting rat basophilic leukemia cells, *Nature (London),* 313, 59, 1985.
48. **Katakami, Y., Kaibuchi, K., Sawamura, M., Takai, Y., and Nishizuka, Y.,** Synergistic action of protein kinase C and calcium for histamine release from rat peritoneal mast cells, *Biochem. Biophys. Res. Commun.,* 121, 573, 1984.
49. **Sagi-Eisenberg, R., Foreman, J. C., and Shelly, R.,** Histamine release induced by histone and phorbol ester from rat peritoneal mast cells, *Eur. J. Pharmacol.,* 113, 11, 1985.
50. **Brockelhurst, K. W., Morita, K., and Pollard, H. B.,** Characterization of protein kinase C and its role in catecholamine secretion from bovine adrenal-medullary cells, *Biochem. J.,* 228, 35, 1985.
51. **Knight, D. E. and Baker, P. F.,** The phorbol ester TPA increases the affinity of exocytosis for calcium in "leaky" adrenal medullary cells, *FEBS Lett.,* 160, 98, 1983.
52. **Frye, R. A. and Holz, R. W.,** Arachidonic acid release and catecholamine secretion from digitonin treated chromaffin cells: effects of micromolar calcium, phorbolester, and protein alkylating agents, *J. Neurochem.,* 44, 265, 1985.
53. **Pocotte, S. L., Frye, R. A., Senter, R. A., TerBush, D. R., Lee, S. A., and Holz, R. W.,** Effects of phorbol ester on catecholamine secretion and protein phosphorylation in adrenal medullary cell cultures, *Proc. Natl. Acad. Sci. U.S.A.,* 82, 930, 1985.
54. **Kojima, I., Kojima, K., Kreutter, D., and Rasmussen, H.,** The temporal integration of the aldosterone secretory response to angiotensin occurs via two intracellular pathways, *J. Biol. Chem.,* 259, 14448, 1984.
55. **Tamagawa, T., Niki, H., and Niki, A.,** Insulin release independent of a rise in cytosolic free Ca^{2+} by forskolin and phorbol ester, *FEBS Lett.,* 183, 430, 1985.
56. **Jones, P. M., Stutchfield, J., and Howell, S. L.,** Effects of Ca^{2+} and a phorbol ester on insulin secretion from islets of Langerhans permeabilised by high-voltage discharge, *FEBS Lett.,* 191, 102, 1985.
57. **Malaisse, W. J., Dunlop, M. E., Mathias, P. C., Malaisse-Lagae, F., and Sener, A.,** Stimulation of protein kinase C and insulin release by 1-oleoyl-2-acetyl-glycerol, *Eur. J. Biochem.,* 149, 23, 1985.
58. **Zawalich, W., Brown, C., and Rasmussen, H.,** Insulin secretion: combined effect of phorbol ester and A23187, *Biochem. Biophys. Res. Commun.,* 117, 448, 1983.
59. **Tanigawa, K., Kuzuya, H., Imura, H., Taniguchi, H., Baba, S., Takai, Y., and Nishizuka, Y.,** Calcium-activated, phospholipid-dependent protein kinase in rat pancreatic islets of Langerhans. Its possible role in glucose-induced insulin release, *FEBS Lett.,* 138, 183, 1982.
60. **Hii, C. S., Stutchfield, J., and Howell, S. L.,** Enhancement of glucagon secretion from isolated rat islets of Langerhans by phorbol 12-myristate 13-acetate, *Biochem. J.,* 233, 287, 1986.
61. **Hutton, J. C., Peshavaria, M., and Brocklehurst, K. W.,** Phorbol ester stimulation of insulin release and secretory-granule protein phosphorylation in a transplantable rat insulinoma, *Biochem. J.,* 224, 483, 1984.
62. **Negro-Vilar, A. and Lapetina, E. G.,** 1,2-Didecanoylglycerol and phorbol 12,13-dibutyrate enhance anterior pituitary hormone secretion *in vitro, Endocrinology,* 117, 1559, 1985.
63. **Ohmura, E. and Friesen, H. G.,** 12-O-Tetradecanoyl phorbol-13-acetate stimulate rat growth hormone (GH) release through different pathways from that of human pancreactic GH-releasing factor, *Endocrinology,* 116, 728, 1985.
64. **Harris, C. E., Staley, D., and Conn, P. M.,** Diacylglycerols and protein kinase C. Potential amplifying mechanism for Ca^{2+}-mediated gonadotropin-releasing hormone-stimulated luteinizing hormone release, *Mol. Pharmacol.,* 27, 532, 1985.
65. **Judd, A. M., Kioke, K., Yasumoto, Y., and MacLeod, R. M.,** Protein kinase C activators and calcium-mobilizing agents synergistically increase GH, LH, and TSH secretion from anterior pituitary cells, *Neuroendocrinology,* 42, 197, 1986.
66. **Chang, J. P., Graeter, J., and Catt, K. J.,** Coordinate action of arachdonic acid and protein kinase C in gonadotropin-releasing hormone-stimulated secretion of leuteinizing hormone, *Biochem. Biophys. Res. Commun.,* 134, 134, 1986.
67. **Conn, P. M., Ganong, B. R., Ebeling, J., Staley, D., Niedel, J. E., and Bell, R. M.,** Diacylglycerols release LH: structure-activity relations reveal a role for protein kinase C, *Biochem. Biophys. Res. Commun.,* 126, 532, 1985.
68. **Naor, Z. and Eli, Y.,** Synergistic stimulation of luteinizing hormone (LH) release by protein kinase C activators and Ca^{2+}-ionophore, *Biochem. Biophys. Res. Commun.,* 130, 848, 1985.
69. **Ronning, S. A. and Martin, T. F.,** Prolactin secretion in permeable GH3 pituitary cells is stimulated by Ca^{2+} and protein kinase C activators, *Biochem. Biophys. Res. Commun.,* 130, 524, 1985.
70. **Delbeke, D., Kojima, I., Dannies, P. S., and Rasmussen, H.,** Synergistic stimulation of prolactin release by phorbol ester, A23187 and forskolin, *Biochem. Biophys. Res. Commun.,* 123, 735, 1984.
71. **Abou-Samra, A. B., Catt, K. J., and Aguilera, G.,** Involvement of protein kinase C in the regulation of adrenocortincotropin release from rat anterior pituitary cells, *Endocrinology,* 118, 212, 1986.

72. **Martin, T. F. and Kowalchyk, J. A.,** Evidence for the role of calcium and diacylglycerol as dual second messengers in thyrotropin-releasing hormone action: involvement of diacylglycerol, *Endocrinology,* 115, 1517, 1984.

73. **Muff, R. and Fischer, J. A.,** Stimulation of parathyroid hormone secretion by phorbol esters is associated with a decrease of cytosolic calcium, *FEBS Lett.,* 194, 215, 1986.

74. **Nemeth, E. F., Wallace, J., and Scarpa, A.,** Stimulus-secretion coupling in bovine parathroid cells. Dissociation between secretion and net change in cytosolic Ca^{2+}, *J. Biol. Chem.,* 261, 2668, 1986.

75. **Brown, E. M., Redgrave, J., and Thatcher, J.,** Effect of the phorbol ester TPA on PTH secretion. Evidence for a role for protein kinase C in the control of PTH release, *FEBS Lett.,* 175, 72, 1984.

76. **Hishikawa, R., Fukase, M., Yamatani, T., Kadowaki, S., and Fujita, T.,** Phorbol ester stimulates calcitonin secretion synergistically with A23187, and additively with dibutyryl cyclic AMP in a rat C-cell line, *Biochem. Biophys. Res. Commun.,* 132, 424, 1985.

77. **de Pont, J. J. H. H. M. and Fleuren-Jakobs, A. M. M.,** Synergistic effect of A23187 and a phorbol ester on amylase secretion from rabbit pancreatic acini, *FEBS Lett.,* 170, 64, 1984.

78. **Merritt, J. E. and Rubin, R. P.,** Pancreatic amylase secretion and cytoplasmic free calcium. Effect of ionomycin, phorbol dibutyrate and diacylglycerols alone and in combination, *Biochem. J.,* 230, 151, 1985.

79. **Putney, J. W. Jr., McKinney, J. S., Aub, D. L., and Leslie, B. A.,** Phorbol ester-induced protein secretion in rat parotid gland. Relationship to the role of inositol lipid breakdown and protein kinase C activation in stimulus-secretion coupling, *Mol. Pharmacol.,* 26, 261, 1984.

80. **Sakamoto, C., Matozaki, T., Nagao, M., and Baba, S.,** Combined effect of phorbol ester and A23187 or dibutyryl cyclic AMP on pepsinogen secretion from isolated gastric gland, *Biochem. Biophys. Res. Commun.,* 131, 314, 1985.

81. **Anderson, N. G. and Hanson, P. J.,** Involvement of calcium sensitive phospholipid-dependent protein kinase in control of acid secretion by isolated rat parietal cells, *Biochem. J.,* 232, 609, 1985.

82. **Sano, K., Voelker, D. R., and Mason, R. J.,** Involvement of protein kinase C in pulmonary surfactant secretion from alveolar type II cells, *J. Biol. Chem.,* 260, 12725, 1985.

83. **Tanaka, C., Taniyama, K., and Kusunoki, M.,** A phorbol ester and A23187 act synergistically to release acetylcholine from the guinea pig ileum, *FEBS Lett.,* 175, 165, 1984.

84. **Publicover, S. J.,** Stimulation of spontaneous transmitter release by the phorbol ester, 12-O-tetradeca-noylphorbol-13-acetate, an activator of protein kinase C, *Brain Res.,* 333, 185, 1985.

85. **Tanaka, C., Fujiwara, H., and Fujii, Y.,** Acetylcholine release from guinea pig caudate slices evoked by phorbol ester and calcium, *FEBS Lett.,* 195, 129, 1986.

86. **Pozzan, T., Gatti, G., Dozio, N., Vicentini, L. M., and Meldolesi, J.,** Ca^{2+}-dependent and -independent release of neurotransmitters from PC12 cells: a role for protein kinase C activation? *J. Cell Biol.,* 99, 628, 1984.

87. **Zurgil, N. and Zisapel, N.,** Phorbol ester and calcium act synergistically to enhance neurotransmitter release by brain neurons in culture, *FEBS Lett.,* 185, 257, 1985.

88. **Rittenhouse, S. E. and Sasson, J. P.,** Mass changes in myoinositol trisphosphate in human platelets stimulated by thrombin, *J. Biol. Chem.,* 260, 8657, 1985.

89. **MacIntyre, D. E., McNicol, A., and Drummond, A. H.,** Tumour-promoting phorbol esters inhibit agonist-induced phosphatidate formation and Ca^{2+} flux in human platelets, *FEBS Lett.,* 180, 160, 1985.

90. **Watson, S. P. and Lapetina, E. G.,** 1,2-Diacylglycerol and phorbol ester inhibit agonist-induced formation of inositol phosphate in human platelets: possible implications for negative feedback regulation of inositol phospholipid hydrolysis, *Proc. Natl. Acad. Sci. U.S.A.,* 82, 2623, 1985.

91. **Zavoico, G. B., Halenda, S. P., Sha'afi, R. I., and Feinstein, M. B.,** Phorbol myristate acetate inhibits thrombin-stimulated Ca^{2+} mobilization and phosphatidylinositol 4,5-bisphosphate hydrolysis in human plate-lets, *Proc. Natl. Acad. Sci. U.S.A.,* 82, 3859, 1985.

92. **Naccache, P. H., Molski, T. F. P., Borgeat, P., White, J. R., and Sha'afi, R. I.,** Phorbol esters inhibit the fMet-Leu-Phe- and leukotriene B₄-stimulated calcium mobilization and enzyme secretion in rabbit neutrophils, *J. Biol. Chem.,* 260, 2125, 1985.

93. **Vicentini, L. M., di Virgilo, F., Ambrosini, A., Pozzan, T., and Meldolesi, J.,** Tumor promoter phorbol 12-myristate, 13-acetate inhibits phosphoinositide hydrolysis and cytosolic Ca^{2+} rise induced by the acti-vation of muscarinic receptor in PC12 cells, *Biochem. Biophys. Res. Commun.,* 127, 310, 1985.

94. **Cooper, R. H., Coll, K. E., and Williamson, J. R.,** Differential effects of phorbol ester on phenylephrine and vasopressin-induced Ca^{2+} mobilization in isolated hepatocytes, *J. Biol. Chem.,* 260, 3281, 1985.

95. **Lynch, C. J., Charest, R., Bocckino, S. B., Exton, J. H., and Blackmore, P. F.,** Inhibition of hepatic α_1-adrenergic effects and binding by phorbol myristate acetate, *J. Biol. Chem.,* 260, 2844, 1985.

96. **Corvera, S. and Garcia-Sainz, J. A.,** Phorbol esters inhibit α_1-adrenergic stimulation of glycogenolysis in isolated rat hepatocytes, *Biochem. Biophys. Res. Commun.,* 119, 1128, 1984.

97. **Orellana, S. A., Solski, P. A., and Brown, J. H.,** Phorbol ester inhibits phosphoinositide hydrolysis and calcium mobilization in cultured astrocytoma cells, *J. Biol. Chem.,* 260, 5236, 1985.

98. **Labarca, R., Janowsky, A., Patel, J., and Paul, S. M.,** Phorbol ester inhibits agonist-induced [³H]inositol-1-phosphate accumulation in rat hippocampal slices, *Biochem. Biophys. Res. Commun.,* 123, 703, 1984.

99. **Drummond, A. H.,** Bidirectional control of cytosolic free calcium by thyrotropin-releasing hormone in pituitary cells, *Nature (London),* 315, 752, 1985.

100. **Tsien, R. Y., Pozzan, T., and Rink, T. J.,** T-cell mitogens cause early changes in cytoplasmic free Ca²⁺ and membrane potential in lymphocytes, *Nature (London),* 295, 68, 1982.

101. **Moolenaar, W. H., Tertoolen, L. G. J., and de Laat, S. W.,** Phorbol ester and diacylglycerol mimic growth factors in raising cytoplasmic pH, *Nature (London),* 312, 371, 1984.

102. **Albert, P. R. and Tashjian, A. H.,** Dual actions of phorbol esters on cytosolic free Ca²⁺ concentrations and reconstitution with ionomycin of acute thyrotropin-releasing hormone responses, *J. Biol. Chem.,* 260, 8746, 1985.

103. **Lagast, H., Pozzan, T., Waldvogel, F. A., and Lew, P. D.,** Phorbol myristate acetate stimulates ATP-dependent calcium transport by the plasma membrane of neutrophils, *J. Clin. Invest.,* 73, 878, 1984.

104. **Kelleher, D. J., Pessin, J. E., Ruoho, A. E., and Johnson, G. L.,** Phorbol ester induces desensitization of adenylate cyclase and phosphorylation of the β-adrenergic receptor in turkey erythrocytes, *Proc. Natl. Acad. Sci. U.S.A.,* 81, 4316, 1984.

105. **Sibley, D. R., Nambi, P., Peter, J. R., and Lefkowitz, R. J.,** Phorbol diesters promote β-adrenergic receptor phosphorylation and adenylate cyclase desensitization in duck erythrocytes, *Biochem. Biophys. Res. Commun.,* 121, 973, 1984.

106. **Mukhopadhyay, A. K. and Schumacher, M.,** Inhibition of hCG-stimulated adenylate cyclase in purified mouse Leydig cells by the phorbol ester PMA, *FEBS Lett.,* 187, 56, 1985.

107. **Papadojoulos, V., Carreau, S., and Drosdowsky, M. A.,** Effect of phorbol ester and phospholipase C on LH-stimulated steroidogenesis in purified rat Leydig cells, *FEBS Lett.,* 188, 312, 1985.

108. **Rebois, R. V. and Patel, J.,** Phorbol ester causes desensitization of gonadotropin-responsive adenylate cyclase in a murine Leydig tumor cell line, *J. Biol. Chem.,* 260, 8026, 1985.

109. **Zatz, M.,** Phorbol esters mimic α-adrenergic potentiation of serotonin N-acetyltransferase induction in the rat pineal, *J. Neurochem.,* 45, 637, 1985.

110. **Cronin, M. J. and Canonico, P. L.,** Tumor promoters enhance basal and growth hormone releasing factor stimulated cyclic AMP levels in anterior pituitary cells, *Biochem. Biophys. Res. Commun.,* 129, 404, 1985.

111. **Quilliam, L. A., Dobson, P. R., and Brown, B. L.,** Modulation of cyclic AMP accumulation in GH3 cells by a phorbol ester and thyroliberin, *Biochem. Biophys. Res. Commun.,* 129, 898, 1985.

Chapter 11

REDOX POTENTIAL

W. J. Malaisse and A. Sener

TABLE OF CONTENTS

I. INTRODUCTION

As amply testified in other chapters of this text, the stimulation of secretion in various cell types coincides with simultaneous changes in a number of biochemical and biophysical variables. One of these changes, to be considered in the present chapter, concerns the redox potential. Basically, three sets of information need to be discussed, namely the evidence that stimulation of secretion may coincide with a change in redox potential, its determination, and its possible relevance to stimulus-secretion coupling.

We intend to examine these three themes using the process of nutrient-induced insulin release in the pancreatic B-cell as the leading example, with emphasis on recent findings. This choice is motivated, in part, by the fact that the redox potential was often considered as a key messenger system coupling the metabolism of insulinotropic nutrients in the islet cells to more distal events in the insulin secretion sequence.[1,2] It should be realized, however, that the concepts under consideration in this chapter may apply, *mutatis mutandis*, to the secretory process in several other cell types.

In considering the participation of the redox potential in secretion, the fundamental assumption will be made that the capacity of nutrient secretagogues to stimulate insulin release reflects their capacity to augment oxidative fluxes and increase O_2 consumption in the pancreatic B-cell. This fuel concept was previously reviewed.[3,4]

II. MULTIFACTORIAL COUPLING OF METABOLIC TO SECRETORY EVENTS

Although emphasis here will be given to the role of the redox potential in stimulus-secretion coupling, we wish first to underline our opinion that, in the process of nutrient-stimulated insulin release, the coupling between metabolic events, looked upon as the site of identification of nutrient secretagogues, and more distal events in the secretory sequence, such as the remodeling of ionic fluxes across the plasma membrane and within the islet cells, may well represent a multifactorial mechanism.[5] This view is based, *inter alia*, on the knowledge that distinct functional variables display vastly different dose-action relationships at increasing concentrations of a given nutrient secretagogue, e.g., D-glucose.[3,5] It could be argued, however, that such differences in dose-response relationships merely reflect the existence of stepwise cause-to-action links between distinct events, with a progressive transfer of information between early and sensitive (high affinity) and distal (low affinity) events. For instance, such a transfer process could involve the participation of threshold phenomena, so that an early variable would need to exceed a critical value in order to cause sizeable changes in a distant variable.

A further finding in support of the multifactorial coupling concept was incidentally collected in the following experiments. It is theoretically not possible to maintain both normal extracellular pH and pCO_2 at variable concentrations of HCO_3^-. Nevertheless, in order to reevaluate the possible influence of HCO_3^- upon insulin release,[6,7] islets were incubated for 90 min in media containing (initial concentrations) HCO_3^- of 4 or 24 mM, respectively. A Hepes-NaOH buffer (10 mM; pH 7.4) was also incorporated in the medium, which was equilibrated against a mixture of CO_2/O_2 (1/19, v/v). The results of these experiments are illustrated in Figure 1. They indicate that, at D-glucose concentrations up to 8.3 mM, the integrated value for insulin release was not affected by the nominal change in HCO_3^- concentration. At higher D-glucose levels, however, much less insulin was released at low than normal HCO_3^- concentration. It should be emphasized that the data illustrated in Figure 1 do not inform on the time course of insulin release. With this reservation in mind, the data suggest that the factors responsible for initiation of the secretory response to D-glucose, at intermediate concentrations of the hexose, are not identical to those required for an optimal

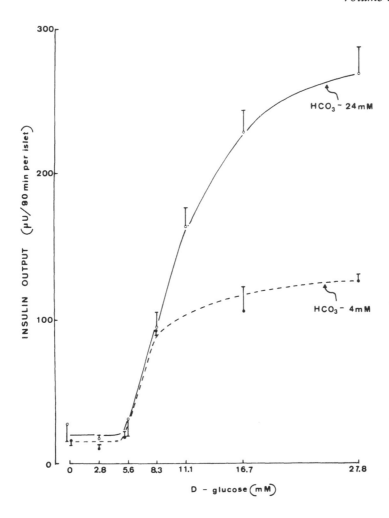

FIGURE 1. Insulin release by islets incubated for 30 min in a Hepes-NaOH buffer (10 mM pH 7.4) containing (initial concentration) 4 or 24 mM HCO_3^- and equilibrated against CO_2/O_2 (1/19, v/v). Mean values ($\pm$ SEM) refer to 12 or more individual observations.

secretory activity at higher D-glucose levels. This again suggests that the process of glucose-induced insulin release may involve a multifactorial stimulus-secretion coupling, the relative contribution of each coupling factor being itself a function of the ambient D-glucose concentration.

The messengers conceivably participating in the multifactorial coupling of metabolic to distal events may include, in addition to the redox potential, proton (H^+) and ATP generated through the metabolism of nutrients, not to mention the possible role of certain metabolites in given catabolic pathways.[8]

III. EFFECT OF NUTRIENT SECRETAGOGUES UPON REDOX POTENTIAL

There is an ample body of evidence indicating that nutrient secretagogues (e.g., D-glucose) cause rapid and sustained changes in the redox potential of pancreatic islet cells.

A. Pyridine Nucleotides
Panten and his colleagues have first shown that glucose and other nutrient secretagogues

augment rapidly the NAD(P)H-fluorescence and decrease the FAD-fluorescence of perifused islets isolated from *ob-ob* mice.[9,10] The latter decrease was interpreted as indicative of a rapid and sustained change in B-cell mitochondrial redox state. However, MacDonald observed that a large fraction of islet flavin was cytosolic rather than mitochondrial.[11]

Further studies have documented that both the NADH/NAD$^+$ and NADPH/NADP$^+$ ratio increase in a rapid and sustained fashion in glucose-stimulated islets.[5,12-14] This was extended to other nutrient secretagogues, and the impression was on occasion gained that the NADPH/NADP$^+$ rather than NADH/NAD$^+$ ratio correlated most closely with the secretory data.[15]

B. Cytosolic Redox Couples

The ratio of reduced to oxidized pyridine nucleotides does not inform on the respective state of the cytosolic and mitochondrial compartments, and does not specifically refer to the cell content in free nucleotides. Therefore, the islet content in critical metabolites was measured to assess specifically the cytosolic redox state of either the NAD or NADP system.

Ashcroft and Christie[16] were the first to show that a rise in D-glucose concentration from 2 to 20 mM increases the cytosolic NADPH/NADP$^+$ ratio in rat pancreatic islets as judged from their malate/pyruvate content. This was confirmed by using either the malate/pyruvate[17] or isocitrate/2-oxoglutarate islet content,[18] and extended to the situation found in islets exposed to 2-ketoisocaproate[17] or L-leucine.[18] D-Glucose and L-leucine also increase the cytosolic NADH/NAD$^+$ ratio both in normal and tumoral islet cells, as judged from the malate/oxalacetate content.[19,20]

C. Glutathione

An increase in the extracellular concentration of D-glucose or other nutrient secretagogues results in an increase in the GSH/GSSG ratio in pancreatic islets.[21,22] According to Ammon et al.,[21] glucose (16.7 mM) increased such a ratio from a basal value of 1.0 to 3.3. Anjaneyulu et al.[22] also observed a modest effect of glucose to increase the islet GSH/GSSG ratio from a basal value of 6.7 to a value of 8.4 at high glucose concentration. The latter data are higher than those reported by Ammon et al.[21] and, as such, closer to values found in other tissues.

IV. MECHANISM OF CHANGES IN REDOX POTENTIAL

A. Generation of Reducing Equivalents

The most naive view concerning the mechanism by which nutrient secretagogues affect redox potential in islet cells would be to postulate that the increased rate of generation of reducing equivalents, resulting from the increase in nutrient oxidation, leads to a rise in the islet content of such equivalents. The magnitude and time course of the changes in redox potential would then be directly dependent on the extent and chronology of the changes in nutrient catabolism.

It should be realized, however, that some non-nutrient secretagogues also affect, to a modest but significant extent, the redox potential in islet cells.[20,23-25] This could suggest that such a potential is not regulated solely by the rate of nutrient oxidation. A note of caution, however, is required here. It is indeed conceivable that secretagogues exerting a primary effect upon biochemical or biophysical events other than nutrient catabolism affect the redox potential through a secondary change in the metabolism of endogenous and/or exogenous nutrients. For instance, it was documented on several occasions that a change in the extracellular ionic environment affects the metabolism of nutrients in islet cells, including changes in O_2 consumption.[26,27] It was even proposed that a feedback control of ionic movements upon nutrient metabolism may play a role in the rhythmicity of bioelectrical and secretory events in the insulin-producing B-cell.[28]

B. Intracellular Distribution of Reducing Equivalents

Reducing equivalents may be formed in the cytosol and/or mitochondria, depending on the nutrient under consideration. The question comes obviously to mind, therefore, whether the transfer of reducing equivalents between these two compartments plays a critical role in the functional response of the islet cells. Only limited information is available on this matter. MacDonald has drawn attention to the possible participation of the glycerol phosphate shuttle[29] and the malate aspartate shuttle[30] in the transfer of reducing equivalents into mitochondria of glucose-stimulated islets, as to maintain a low NADH/NAD$^+$ cytosolic ratio and, hence, a sustained rate of glycolysis. The participation of the glycerol phosphate shuttle will be discussed later in this report. The possible participation of the malate aspartate shuttle was examined indirectly by investigating the effect of aminooxyacetate, an inhibitor of aminotransferase, upon several metabolic, ionic, and secretory variables in pancreatic islets.[17,31] It was observed that aminooxyacetate severely affects the cationic and secretory response of the islets to various nutrient secretagogues, including D-glucose and 2-ketoisocaproate, without causing any obvious alteration in the oxidation of these nutrients. It remains, however, to document unambiguously in which direction the shuttle operates in islets exposed to distinct nutrient secretagogues.

In a somewhat parallel perspective, it should be underlined that the insulinotropic action of certain nutrient secretagogues may be due to an accelerated generation rate of cytosolic NADPH, rather than to any sizeable increase in islet respiration. This appears to be the case, for instance, in islets exposed to L-asparagine.[32-34]

C. Reoxidation of Reducing Equivalents

In addition to being affected by changes in the rate of generation and intracellular distribution of reducing equivalents, it is conceivable that the steady-state redox potential of the NADH/NAD$^+$ and NADPH/NADP$^+$ systems, in either the cytosol or mitochondria, is modulated by the rate of reoxidation of these equivalents. This is an essential issue since it directly concerns the link between the generation of reducing equivalents and ATP, respectively. Once again, however, only limited information is presently available, in pancreatic islet cells, on this specific issue. For instance, since nutrient secretagogues increase both the GSH/GSSG ratio and the rate of fatty acid synthesis[32,35,36] in the islets, the consumption of NADPH in either the reaction catalyzed by glutathione reductase[22] or synthesis of acyl residues should be assessed and duly taken into account to establish the balance between the generation and consumption of reducing equivalents. To further illustrate our point, we would like to refer to a recent study on the modality of cytosolic NADH reoxidation in both normal and tumoral islet cells stimulated by D-glucose. D-Glucose catabolism was examined in rat pancreatic islets or tumoral insulin-producing cells of the RINm5F line exposed to 2.8 or 16.7 mM D[3,4-^{14}C]glucose. The production of labeled lactate from D-[3,4-^{14}C]glucose was judged by the lactate oxidase method. The total utilization of D-[3,4-^{14}C]glucose was taken as the sum of the formation of $^{14}CO_2$ and labeled lactate, pyruvate, and alanine. The circulation in the glycerol phosphate shuttle was estimated from the utilization of D-[3,4-^{14}C]glucose unaccounted for by the production of [^{14}C]lactate. As indicated in Table 1, the flow rate in the glycerol phosphate shuttle was four times higher in islets exposed to 16.7 mM rather than 2.8 mM D-glucose, while being unaffected by the hexose concentration in RINm5F cells. Even when expressed relative to the total rate of D-[3,4-^{14}C]glucose utilization, the flow rate through the glycerol phosphate shuttle remained higher ($p < 0.005$) in the islets exposed to 16.7 mM D-glucose than in those incubated in the presence of 2.8 mM D-glucose. This was not the case in the RINm5F cells. This and other findings indicated that normal, but not tumoral, islet cells are organized to favor those mitochondrial reactions yielding the major fraction of the total energy generated by the catabolism of D-glucose.[37]

Table 1

D-[3,4-¹⁴C]GLUCOSE METABOLISM IN RAT PANCREATIC
ISLETS AND TUMORAL INSULIN-PRODUCING CELLS
(RINm5F LINE)

Tissue:	Pancreatic islets		RINm5F cells	
D-Glucose (m*M*):	2.8	16.7	2.8	16.7

D-[3,4-¹⁴C]glucose utilization (pmol/120 min/islet or 10³ cells)

	72.5 ± 5.8	228.2 ± 7.5	142.8 ± 18.1	189.2 ± 5.8

[¹⁴C]lactate production (pmol/120 min/islet or 10³ cells)

	52.4 ± 5.7	137.2 ± 6.1	110.5 ± 18.0	152.6 ± 5.6

Flow through the glycerol phosphate shuttle (pmol/120 min/islet or 10³ cells)

	20.1 ± 1.3	91.0 ± 4.3	32.3 ± 1.6	36.6 ± 1.5

Flow through the glycerol phosphate shuttle/D-[3,4-¹⁴C]glucose utilization (%)

	27.7 ± 2.9	39.9 ± 2.3	22.6 ± 3.1	19.3 ± 1.0

V. TARGET SYSTEMS FOR CHANGES IN REDOX POTENTIAL

By which mechanism does a change in redox potential eventually lead to a change in secretory rate? Or, in other words, which systems act as targets responsive to the redox potential? The most naive answer could be that by affecting the thiol-disulfide balance of selected proteins, the change in redox potential modifies the participation of these proteins in suitable regulatory processes. A prerequisite of the latter postulate is that nutrient secretagogues indeed modify the content of thiol groups in the islet cells.

A. Thiol Groups

Histochemical detection of SH-groups in pancreatic islets demonstrates a lower content than in the exocrine pancreas.[38] D-Glucose (16.7 m*M*) augments the islet content in thiol groups from a basal value of 119 ± 7 to 170 ± 9 pmol/µg protein, as measured by the 5-5'-dithiobis(2-nitrobenzoate) technique.[22] Other nutrient secretagogues duplicate the effect of D-glucose.

Little is known of the enzymes involved in the control of thiol:disulfide balance in pancreatic islets. Islet homogenates display, in addition to glutathione reductase activity, little glutathione peroxidase activity but high glutathion-cysteine transhydrogenase activity.[22] Täljedal[39] also identified thioredoxin in pancreatic islets and proposed that it may participate in electron transport to the B-cell plasma membrane.

B. Enzymes Affected by Changes in Redox Potential

Several enzymes involved in stimulus-secretion coupling are susceptible to undergo changes in activity as a result of a change in redox potential. We will restrict the illustration of such a concept to three examples.

Glyceraldehyde 3-phosphate dehydrogenase is susceptible to inhibition by reagents reacting with thiols. This provided a tool for early investigations on the relevance of glycolysis to the process of glucose-stimulated insulin release.[40] It is obvious, however, that the regulation of glycolysis and several other metabolic pathways may be affected not solely by a change in the thiol-disulfide balance of relevant enzymes but also by the availability of oxidized or reduced pyridine nucleotides.

Capito et al. observed that NADH inhibits Ca-ATPase activity in a granule fraction prepared from obese mice islets. The inhibitory effect of NADH was greater than that of NAD⁺ and could not be attributed to chelation of Ca²⁺. It was postulated that NADH may increase cytosolic free Ca²⁺ by inhibiting Ca-ATPase and thereby decreasing the rate at which calcium is removed from the cytosol.[41]

Recent studies have led to the concept that transglutaminase may participate in motile events involved in both the conversion of proinsulin to insulin and the release of insulin by exocytosis.[42,43] This enzyme operates by a cysteine thiol-active center mechanism. It was proposed that activation of the enzyme in glucose-stimulated islets may be attributable not solely to an increase in cytosolic Ca^{2+} activity but also to the induction of a more reduced redox state with subsequent change in thiol-disulfide balance.[44]

C. Other Possible Targets

In addition to enzymes, other systems involved in the stimulus-secretion coupling process might be modulated by changes in redox potential. Three examples will again be mentioned.

Hellman and his colleagues[45] were the first to propose that ionic fluxes across the B-cell plasma membrane are regulated by proteins with prevalent disulfide bridges in the resting state and prevalent sulfhydryl groups in the stimulated state. Later, when it became evident that a decrease in K^+ conductance may play a critical role in the secretory sequence, the view was expressed that a change in redox potential might control the cell membrane permeability to K^+.[1,46] More recently, however, electrophysiological studies have emphasized the possible role of pH-sensitive and/or ATP-sensitive K^+ channels in the control of K^+ conductance.[47-50]

Second, by analogy with the situation described in other tissues,[51] it was proposed that a change in mitochondrial redox states may play a critical role in the fluxes of Ca^{2+} between the mitochondrial and cytosolic compartment.[52] Thus, according to Lehninger et al., a more reduced state of mitochondrial pyridine nucleotides could promote Ca^{2+} uptake by mitochondria.[53]

Third, to the extent that native ionophores participate in the transport of such ions as H^+, Na^+, and Ca^{2+} across membrane systems,[54-57] it is conceivable that a change in redox potential affects these ionophoretic processes. For instance, the ionophoretic properties of X537A are different from those of its dihydroderivative.[3] Furthermore, GSH and GSSG affect differently the ionophoretic potential of both X537A and an islet lipid extract.[58]

These examples are not exhaustive and merely indicate that there is no shortage of potential candidates as target systems affected by changes in redox potential.

VI. DRUG-INDUCED INTERFERENCES

Apart from the information gained from the direct measurement of redox variables in cells exposed to physiological stimuli, the possible relevance of changes in redox state to the process of stimulus-secretion coupling can also be explored indirectly by using drugs supposed to alter, in a site-specific manner, reactions involved in such changes. This approach usually meets with considerable limitations in that the alleged site specificity may always be questioned. In order to illustrate such a point, we will consider three exemplative situations.

A. Interference with NAD(P)H Availability

In an attempt to document the role of NAD(P)H in nutrient-stimulated insulin release, we had examined the effect of menadione upon metabolic, cationic, and secretory variables in isolated pancreatic islets.[1,2] Pancreatic islets contain the enzyme system which catalyzes the donation of hydrogen from NAD(P)H to menadione. In high concentrations (20 to 50 μM), menadione, in addition to lowering the concentration of reduced pyridine nucleotide in the islets, also impaired glycolysis and glucose oxidation, decreased ATP content, and inhibited proinsulin biosynthesis. However, in the presence of 10 μM menadione and 11.1 mM D-glucose, menadione failed to affect the concentration of adenine nucleotides, the utilization of D-[5-³H]glucose, the production of lactate and pyruvate, the oxidation of D-[6-¹⁴C]glucose, and the synthesis of proinsulin; whereas the metabolism of D-glucose through the pentose

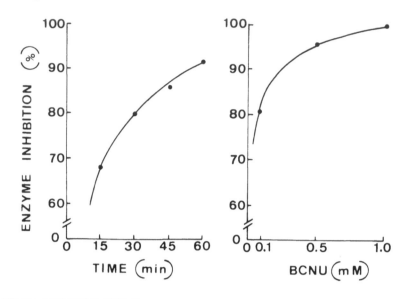

FIGURE 2.   Time course (left) and dose-action relationship (right) for
the inhibitory action of BCNU upon the activity of glutathione reductase.
The islets were incubated for 15 to 60 min in the presence of 0.5 mM
BCNU (left) or for 60 min in the presence of 0.1 to 1.0 mM BCNU (right),
prior to being homogenized for measurement of enzyme activity. Mean
values refer to 2 or more individual measurements, and are expressed in
percent of the paired control value found in the absence of BCNU.

shunt was markedly increased. The sole inhibitory effect of menadione (10 μM) upon metabolic parameters was to reduce the concentration of both NADH and NADPH. In the
presence of 11.1 mMD-glucose, menadione (10 μM) decreased ^{45}Ca uptake and insulin
release, this coinciding with a delayed inhibition by D-glucose of ^{86}Rb outflow and a decreased magnitude of the secondary rise in ^{45}Ca outflow normally evoked by the hexose in
prelabeled islets. These data were interpreted to indicate that the availability of reduced
pyridine nucleotides plays an essential role in the secretory sequence by coupling metabolic
to cationic events. It should be realized, however, that even if the ATP content of the islet
cells was unaffected by menadione (10 μM), its generation rate was probably decreased.

B. Interference with Glutathione Reductase

The participation of glutathione reductase in the process of nutrient-stimulated insulin
release was investigated in rat pancreatic islets exposed to a selective inhibitor of this enzyme,
1,3-bis(2-chloroethyl)-1-nitrosourea (BCNU).[59] BCNU (0.1 to 1.0 mM) indeed caused a
dose-related inhibition of glutathione reductase. At a fixed concentration of the drug (0.5
mM), the effect was time-related (Figure 2). The enzyme inhibition was not reversible, at
least over the ensuing 60-min incubation. BCNU (0.05 to 1.0 mM) also caused a dose-
related inhibition of insulin release evoked by either D-glucose or the combination of L-
leucine and L-glutamine, while failing to suppress the secretory response to nonnutrient
secretagogues, such as the association of Ba^{2+} and theophylline. The interpretation of the
data obtained in islets exposed to D-glucose was obscured by the fact that BCNU inhibited
D-[U-^{14}C]glucose oxidation. However, up to a concentration of 0.5 mM, BCNU failed to
affect the oxidation of L-[U-^{14}C]-glutamine and L-[U-^{14}C]leucine. Further studies were restricted, therefore, to the situation found in the presence of the two amino acids. BCNU
failed to affect adversely the nutrient-induced increase in cytosolic NADH/NAD$^+$ and NADPH/
NADP$^+$ ratios. As expected, BCNU lowered the basal GSH/GSSG ratio and suppressed the

increase in such a ratio normally provoked by the amino acids. Likewise, BCNU severely decreased the basal thiol content of the islets and abolished the increment in thiol content normally caused by L-leucine and L-glutamine. BCNU failed to affect the basal or amino acid-stimulated production of cAMP, but altered the cationic response of the islets to these exogenous nutrients. Thus, BCNU decreased ^{45}Ca net uptake measured in the presence, but not absence, of L-leucine and L-glutamine. This could not be attributed to a lesser ability of the amino acids to inhibit K^+ conductance (as judged from the decrease in ^{86}Rb$^+$ outflow from prelabeled islets), raising the idea that the inhibitory action of BCNU upon insulin release was attributable, in part at least, to an anomaly in the intracellular distribution of Ca^{2+}, e.g. between mitochondrial (NADPH-dependent) and extramitochondrial (thiol-dependent) compartments. By altering the thiol-disulfide balance, BCNU could also affect, however, the sensitivity to Ca^{2+} of Ca^{2+}-responsive targets, including transglutaminase.[59]

C. Interference with GSH Availability

Several carbamyl compounds, including 2-cyclohexene-1-one (CHX) decrease reduced glutathione levels in liver and other tissues, possibly by acting as a substrate for glutathione S-transferases,[60-62] and were used, therefore, as a tool to assess the role of glutathione in cell activation. Preincubation of pancreatic islets for 30 min with CHX (1.0 mM or more) lowers their GSH content, while failing to affect the GSSG content. CHX also fails to affect NADH/NAD$^+$ and NADPH/NADP$^+$ cytosolic redox couples, and only caused a marginal alteration of the glucose-induced increase in islet thiol content. Nevertheless, CHX severely decreases insulin release evoked by D-glucose and 2-ketoisocaproate. CHX, however, also inhibits insulin release evoked by nonnutrient secretagogues (e.g., gliclazide or Ba^{2+}) and, at least at concentrations up to 1.0 mM, fails to inhibit L-leucine-stimulated insulin output. L-Glutamine protects the B-cell against the inhibitory action of CHX upon D-glucose-induced insulin release. The latter inhibitory action coincides with an impaired oxidation of D-[U-^{14}C]glucose, whereas the oxidation of L-[U-^{14}C]leucine and L-[U-^{14}C]glutamine are little affected by CHX (1.0 mM). CHX severely perturbs the normal relationship between nutrient oxidation, ^{45}Ca net uptake, and insulin release. Assuming that the primary action of CHX is to lower the islet cell GSH content, these data suggest that GSH availability may exert multiple effects in the islet cells, affecting the catabolism of certain nutrients, the coupling of nutrient oxidation to Ca^{2+} handling, and the response to this cation of Ca^{2+}-sensitive targets, e.g., transglutaminase. The latter proposal was supported by the finding that CHX renders islet cells unresponsive to forskolin, theophylline, and cytochalasin B.[63,64]

VII. DOES REDOX POTENTIAL ACT AS SECOND MESSENGER?

The information so far reviewed in this chapter indicates that the redox potential in islet cells is affected by nutrient secretagogues and that a number of targets could conceivably be affected by such a change in redox potential. The effect upon the secretory behavior of the islet cells of drugs known to affect specific components of the redox system supports the view that the functional integrity of this system is required to ensure an optimal secretory response to nutrient secretagogues.

None of these considerations unambiguously demonstrates that the redox potential indeed acts as a second messenger in the process of nutrient-stimulated insulin release. As a matter of fact, two recent observations suggest that the redox potential may not represent the major coupling factor between metabolic and secretory events. These two observations will now be considered in some detail.

A. Dissociation Between Insulin Release and Cytosolic Redox State in Normal Islet Cells

In order to evaluate the possible role of changes in cytosolic redox state, we have recently

Table 2
EFFECT OF NUTRIENT SECRETAGOGUES UPON CYTOSOLIC REDOX
COUPLES AND INSULIN RELEASE

Nutrient (mM)	Nil		D-Glucose (16.7)		2-Ketoisocaproate (10.0)		L-Leucine (10.0)	
Malate (fmol/islet)	736 ± 36	(34)	1188 ± 43	(35)	1985 ± 72	(35)	1121 ± 49	(35)
Pyruvate (fmol/islet)	532 ± 40	(19)	694 ± 41	(19)	752 ± 47	(19)	617 ± 45	(19)
Oxalacetate (fmol/islet)	13.4 ± 0.7	(27)	12.2 ± 0.6	(28)	26.6 ± 1.0	(28)	9.1 ± 0.7	(28)
NADH/NAD$^+$ (ratio × 10^3)	1.87 ± 0.11	(27)	3.43 ± 0.16	(28)	2.61 ± 0.14	(28)	4.28 ± 0.32	(28)
NADPH/NADP$^+$ (ratio)	45.3 ± 2.6	(19)	64.8 ± 3.0	(19)	84.7 ± 4.0	(19)	66.4 ± 3.5	(19)
Insulin output (μU/islet/90 min)	18.1 ± 1.7	(49)	245.0 ± 4.9	(49)	76.9 ± 1.2	(30)	63.2 ± 1.9	(30)

compared the effects of distinct nutrient secretagogues upon both the cytosolic NADH/ NAD$^+$ and NADPH/NADP$^+$ ratios in rat pancreatic islets. In addition to D-glucose, 2-ketoisocaproate and L-leucine were selected for this study because these noncarbohydrate nutrients are potent insulin secretagogues, despite the fact that they are oxidized solely in mitochondria. The islet content in malate, oxalacetate, and pyruvate was measured after 30 min incubation and compared to the rate of insulin release measured over 90 min incubation. The results are summarized in Table 2.

All three nutrients tested augmented both cytosolic redox states ($p < 0.001$ in all cases). There was no obvious parallelism between the effects of distinct nutrients upon the NADH/ NAD$^+$ ratio and NADPH/NADP$^+$ ratio, respectively. For instance, 2-ketoisocaproate augmented more than the two other nutrients the cytosolic NADPH/NADP$^+$ ratio ($p < 0.005$), but augmented less than the two other secretagogues the cytosolic NADH/NAD$^+$ ratio ($p < 0.001$). Relative to basal value ($24.2 ± 2.0 × 10^3$), the cytosolic [NADPH]·[NAD$^+$]/ [NADH]·[NADP$^+$] ratio was increased ($p < 0.01$) by 2-ketoisocaproate (to $32.5 ± 2.3 × 10^3$) and decreased ($p < 0.001$) by L-leucine (to $15.5 ± 1.4 × 10^3$). There was no correlation between the magnitude of changes in cytosolic redox state and insulin release, respectively.[19]

These findings may have several implications. First, they indicate that D-glucose is not the sole nutrient to augment cytosolic redox couples, as could result from an increased flow rate in both the glycolytic pathway (NADH/NAD$^+$ ratio) and pentose cycle (NADPH/ NADP$^+$ ratio). The data obtained with 2-ketoisocaproate and L-leucine suggest that reducing equivalents are transferred from the mitochondria into the cytosol of islet cells exposed to these exogenous nutrients. The transfer of NADH may be mediated by the malate-aspartate shuttle,[17,30] and that of NADPH by the shuttle system involving the cytosolic and mitochondrial NADP-specific isocitrate dehydrogenase, which were both recently identified in islet homogenates.[18]

Second, although the present data refer to cytosolic, rather than mitochondrial, ratios, they may inform on the activity of the energy-linked mitochondrial nicotinamide nucleotide transhydrogenase. For instance, 2-ketoisocaproate could favor the energy-linked transfer of hydrogen from NADH to NADP$^+$, whereas L-leucine could favor the reduction of NAD$^+$ by NADPH. The latter situation may be related to the fact that L-leucine, by activating glutamate dehydrogenase [NAD(P)$^+$], augments the mitochondrial rate of generation of both NADH and NADPH.

Finally, the dissociation between changes in cytosolic redox state and insulin release, respectively, in response to distinct nutrient secretagogues suggests that a more reduced state of cytosolic redox couples does not represent the sole, and possibly even not the major, determinant of nutrient-induced insulin secretion. Such a conclusion is not meant to deny that changes in cytosolic redox potential may nevertheless play a significant, if limited, role in the multifactorial coupling of metabolic to secretory events in the process of nutrient-stimulated insulin release.

B. Dissociation Between Insulin Release and Cytosolic Redox State in Tumoral Islet Cells

The nonmetabolized analog of L-leucine, 2-aminobicyclo-[2,2,1]heptane-2-carboxylic acid (BCH) stimulates insulin release in rat pancreatic islets. This effect is attributable to activation of glutamate dehydrogenase, leading to an accelerated catabolism of endogenous amino acids and an increased consumption of O_2. These metabolic changes result in turn in the stimulation of those ionic and secretory events usually observed in the process of nutrient-stimulated insulin release.[65-69] In the course of investigations on the metabolic and secretory behavior of tumoral insulin-producing cells (RINm5F line), we have recently identified a different response to BCH than that previously observed in normal islet cells.[20]

In RINm5F cell homogenates, like in normal islet cell homogenates, BCH was unable to act as a transamination partner, but activated glutamate dehydrogenase. In intact tumoral cells, like in intact normal islet cells, BCH stimulated the oxidation of exogenous L-[U-^{14}C]glutamine and induced a more reduced state of both the cytosolic NADH/NAD$^+$ and NADPH/NADP$^+$ ratio. Yet, in sharp contrast to its effect in normal rat islets, BCH suppressed the rather elevated basal release of insulin from the RINm5F cells, this coinciding with inhibition of basal O_2 uptake, a fall in the total adenine nucleotide content, ATP/ADP ratio and adenylate charge, and an increased fractional outflow rate of ^{86}Rb from prelabeled RINm5F cells. Although we have not yet identified those endogenous nutrients which may be oxidized at a lower rate in BCH-treated cells, these data again indicate that the induction of a more reduced cytosolic redox state may not be the critical factor regulating insulin release. Instead, the present data are compatible with the view that the rate of ATP generation represents an essential determinant of the secretory response to nutrient secretagogues, as already proposed some years ago.[27] This proposal should be considered in the light of recent electrophysiological work performed with the patch clamp technique and suggesting that ATP-responsive K$^+$ channels may play a key role in the sequence of cationic events involved in the process of nutrient-stimulated insulin release.[48-50]

C. Alternative Coupling Messengers

The experimental data reviewed in this section draw attention to the possible key role of messengers other than a change in redox potential, especially the possible role of cytosolic ATP as a coupling agent between metabolic and selected ionic events in the secretory sequence. In this perspective, we have recently examined the influence of D-glucose upon cytosolic ATP in rat pancreatic islets.[70]

The cytosolic mitochondrial contents in ATP, ADP, and AMP were measured in islets incubated for 45 min at increasing concentrations of D-glucose and then exposed for 20 sec to digitonin. The latter treatment failed to affect the total islet ATP/ADP ratio and adenylate charge. D-Glucose caused a much greater increase in cytosolic than mitochondrial ATP/ADP ratio (Figure 3). In the cytosol, a sigmoidal pattern characterized the changes in ATP/ADP ratio at increasing concentrations of D-glucose. These findings are compatible with the view that cytosolic ATP participates in the coupling of metabolic to ionic events, e.g., by inactivating ATP-sensitive K$^+$ channels.

VIII. CONCLUSIONS

The information reviewed in this chapter may well leave a feeling of unease. On one hand, convincing evidence is produced to indicate that the stimulation of insulin secretion by nutrient secretagogues coincides with a change in the redox potential of islet cells. On the other hand, attention is drawn to a number of uncertainties concerning the regulation and regulatory role(s) of this change in redox potential.

First, the respective contribution of distinct nutrients to the overall generation of reducing

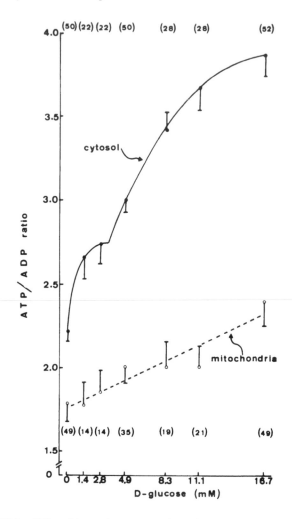

FIGURE 3.   Effect of increasing concentrations of D-glucose upon the cytosolic
and mitochondrial ATP/ADP ratios in rat pancreatic islets. Mean values ($\pm$ SEM)
refer to the number of individual determinations quoted in parentheses.

equivalents and, hence, the subcellular location of such a generation in the islet cells need
to be defined with greater precision. Second, the transfer of reducing equivalents between
the mitochondrial and cytosolic compartments and the regulation of such a transfer remain
poorly understood. Third, the major target systems affected by a change in redox potential
have not yet been well identified or characterized. Finally, the respective role of changes
in redox potential and other coupling factors in the control of functional events in the islet
cells still appears as a matter of debate.

It would seem inappropriate, however, to conclude this review in a pessimistic mood.
Instead, the fact that several essential questions remain unanswered could be taken as a
guarantee for the fruitfulness of further investigations on the significance of redox potential
in secretory processes.

ACKNOWLDGEMENTS

The experimental work reviewed in this chapter was supported in part by grants from the
Belgian Foundation for Scientific Medical Research. The authors are grateful to C. De-
mesmaeker for secretarial assistance.

REFERENCES

1. **Malaisse, W. J., Sener, A., Boschero, A. C., Kawazu, S., Devis, G., and Somers, G.,** The stimulus-secretion coupling of glucose-induced insulin release. Cationic and secretory effects of menadione in the endocrine pancreas, *Eur. J. Biochem.,* 87, 111, 1978.
2. **Malaisse, W. J., Hutton, J. C., Kawazu, S., and Sener, A.,** Metabolic effects of menadione in isolated islets, *Eur. J. Biochem.,* 87, 121, 1978.
3. **Malaisse, W. J., Sener, A., Herchuelz, A., and Hutton, J. C.,** Insulin release: the fuel hypothesis, *Metabolism,* 28, 373, 1979.
4. **Malaisse, W. J.,** Insulin release: the fuel concept, *Diabetes Metab.,* 9, 313, 1983.
5. **Malaisse, W. J., Hutton, J. C., Kawazu, S., Herchuelz, A., Valverde, I., and Sener, A.,** The stimulus-secretion coupling of glucose-induced insulin release. XXXV. The links between metabolic and cationic events, *Diabetologia,* 16, 331, 1979.
6. **Henquin, J. C. and Lambert, A. E.,** Extracellular bicarbonate ions and insulin release, *Biochim. Biophys. Acta,* 381, 437, 1975.
7. **Lebrun, P., Malaisse, W. J., and Herchuelz, A.,** Effect of bicarbonate upon intracellular pH and calcium fluxes in pancreatic islet cells, *Biochem. Biophys. Acta,* 721, 357, 1982.
8. **Malaisse, W. J., Malaisse-Lagae, F., and Sener, A.,** Coupling factors in nutrient-induced insulin release, *Experientia,* 40, 1035, 1984.
9. **Panten, U., Christians, J. v. Kriegstein, E., Poser, W., and Hasselblatt, A.,** Effect of carbohydrates upon fluorescence of reduced pyridine nucleotides from perifused isolated pancreatic islets, *Diabetologia,* 9, 477, 1973.
10. **Panten, U. and Ishida, H.,** Fluorescence of oxidized flavoproteins from perifused isolated pancreatic islets, *Diabetologia,* 11, 569, 1975.
11. **MacDonald, M. J.,** Flavin content of intracellular compartments of pancreatic islets compared with acinar tissue and liver, *Endocrinology,* 108, 1899, 1981.
12. **Ammon, H. P. T.,** Effect of tolbutamide on aminophylline-3,5-AMC-dibutyrate and glucagon-induced insulin release from pancreatic islets after impairment of pyridine nucleotide metabolism caused by 6-aminonicotinamide (6-AN), *Naunyn-Schmiedebergs Arch. Pharmakol.,* 290, 251, 1975.
13. **Brolin, S. E., Ågren, A., and Peterson, B.,** Determinations of redox states in A₂ and B-cell rich islet specimens from guinea pigs, using bioluminescence assay of NAD⁺ and NADH, *Acta Endocrinol.,* 96, 93, 1981.
14. **Deery, D. J. and Taylor, K. W.,** Effects of azaserine and nicotinamide on insulin release and nicotinamide-adenine dinucleotide metabolism in isolated rat islets of Langerhans, *Biochem. J.,* 134, 557, 1973.
15. **Hutton, J. C., Sener, A., and Malaisse, W. J.,** The stimulus-secretion coupling of 4-methyl-2-oxopentanoate-induced insulin release, *Biochem. J.,* 184, 303, 1979.
16. **Ashcroft, S. J. H. and Christie, M. R.,** Effects of glucose on the cytosolic ratio of reduced/oxidized nicotinamide-adenine dinucleotide phosphate in rat islets of Langerhans, *Biochem. J.,* 184, 697, 1979.
17. **Malaisse, W. J., Malaisse-Lagae, F., and Sener, A.,** The stimulus-secretion coupling of glucose-induced insulin release: effect of aminooxyacetate upon nutrient-stimulated insulin secretion, *Endocrinology,* 111, 392, 1982.
18. **Sener, A., Malaisse-Lagae, F., Dufrane, S. P., and Malaisse, W. J.,** The coupling of metabolic to secretory events in pancreatic islets. The cytosolic redox state, *Biochem. J.,* 220, 433, 1984.
19. **Sener, A. and Malaisse, W. J.,** The coupling of metabolic to secretory events in pancreatic islets: dissociation between insulin release and cytosolic redox state, *Biochem. Int.,* 14, 897, 1987.
20. **Sener, A., Leclercq-Meyer, V., Giroix, M.-H., Malaisse, W. J., and Hellerström, C.,** Opposite effects of D-glucose and a nonmetabolized analogue of L-leucine upon respiration and secretion in insulin-producing tumoral cells (RINm5F), *Diabetes,* 36, 187, 1987.
21. **Ammon, H. P. T., Abdel-Hamid, M., Rao, P. G., and Enz, G.,** Thiol-dependent and non-thiol-dependent stimulations of insulin release, *Diabetes,* 33, 251, 1984.
22. **Anjaneyulu, K., Anjaneyulu, R., Sener, A., and Malaisse, W. J.,** The stimulus-secretion coupling of glucose-induced insulin release. Thiol-disulfide balance in pancreatic islets, *Biochimie,* 64, 29, 1982.
23. **Trus, M. D., Hintz, C. S., Weinstein, J. B., Williams, A. D., Pagliara, A. S., and Matschinsky, F. M.,** Effects of glucose and acetycholine on islet tissue NADH and insulin release, *Life Sci.,* 22, 809, 1978.
24. **Sener, A. and Malaisse, W. J.,** The stimulus-secretion coupling of glucose-induced insulin release. Metabolic events in islets stimulated by non-metabolizable secretagogues, *Eur. J. Biochem.,* 98, 141, 1979.
25. **Kawazu, S., Sener, A., Couturier, E., and Malaisse, W. J.,** Metabolic, cationic and secretory effects of hypoglycemic sulfonylureas in pancreatic islets, *Naunyn-Schmiedeberg's Arch. Pharmakol.,* 312, 277, 1980.

26. **Sener, A., Malaisse-Lagae, F., and Malaisse, W. J.,** The stimulus-secretion coupling of glucose-induced insulin release. Environmental influence on L-glutamine oxidation in pancreatic islets, *Biochem. J.,* 202, 309, 1982.

27. **Hutton, J. C. and Malaisse, W. J.,** Dynamics of O_2 consumption in rat pancreatic islets, *Diabetologia,* 18, 395, 1980.

28. **Malaisse, W. J., Sener, A., Herchuelz, A., Valverde, I., Hutton, J. C., Atwater, I., and Leclercq-Meyer, V.,** The interplay between metabolic and cationic events in islet cells: coupling factors and feedback mechanisms, *Horm. Metab. Res.,* Suppl. 10, 61, 1980.

29. **MacDonald, M. J.,** High content of mitochondrial glycerol-3-phosphate dehydrogenase in pancreatic islets and its inhibition by diazoxide, *J. Biol. Chem.,* 256, 8287, 1981.

30. **MacDonald, M. J.,** Evidence for malate aspartate shuttle in pancreatic islets, *Arch. Biochem. Biophys.,* 213, 643, 1982.

31. **Lebrun, P., Malaisse, W. J., and Herchuelz, A.,** Impairment by aminooxyacetate of ionic response to nutrients in pancreatic islets, *Am. J. Physiol.,* 245, E38, 1983.

32. **Malaisse-Lagae, F., Welsh, M., Lebrun, P., Sener, A., Hellerström, C., and Malaisse, W. J.,** The stimulus-secretion coupling of amino acid-induced insulin release: secretory and oxidative response of pancreatic islets to L-asparagine, *Diabetes,* 33, 464, 1984.

33. **Sener, A., Best, L., Malaisse-Lagae, F., and Malaisse, W. J.,** The stimulus-secretion coupling of amino acid-induced insulin release. Metabolism of L-asparagine in pancreatic islets, *Arch. Biochem. Biophys.,* 229, 155, 1984.

34. **Malaisse, W. J., Malaisse-Lagae, F., and Sener, A.,** The stimulus-secretion coupling of amino acid-induced insulin release. Metabolic interaction of L-asparagine and L-leucine in pancreatic islets, *Biochim. Biophys. Acta,* 797, 194, 1984.

35. **Sener, A., Kawazu, S., Hutton, J. C., Boschero, A. C., Devis, G., Somers, G., Herchuelz, A., and Malaisse, W. J.,** The stimulus-secretion coupling of glucose-induced insulin release. Effect of exogenous pyruvate on islet function, *Biochem. J.,* 176, 217, 1978.

36. **Hutton, J. C., Sener, A., and Malaisse, W. J.,** The metabolism of 4-methyl-2-oxopentanoate in rat pancreatic islets, *Biochem. J.,* 184, 291, 1979.

37. **Sener, A. and Malaisse, W. J.,** Stimulation by D-glucose of mitochondrial oxidative events in islet cells, *Biochem. J.,* 246, 89, 1987.

38. **Hahn von Dorsche, H. and Wolter, S.,** The content of SH-groups in the islets of Langerhans, *Acta Histochem.,* 66, 279, 1980.

39. **Täljedal, I.-B.,** On insulin secretion, *Diabetologia,* 21, 1, 1981.

40. **Sener, A., Pipeleers, D. G., Levy, J., and Malaisse, W. J.,** The stimulus-secretion coupling of glucose-induced insulin release. XXVI. Are the secretory and fuel functions of glucose dissociable by iodoacetate?, *Metabolism,* 27, 1505, 1978.

41. **Capito, K., Formby, B., and Hedeskov, C. J.,** Ca-ATPase in pancreatic islets, *Horm. Metab. Res.,* Suppl. 10, 50, 1980.

42. **Sener, A., Dunlop, M. E., Gomis, R., Mathias, P. C. F., Malaisse-Lagae, F., and Malaisse, W. J.,** Role of transglutaminase in insulin release. Study with glycine and sarcosine methylesters, *Endocrinology,* 117, 237, 1985.

43. **Alarcon, C., Valverde, I., and Malaisse, W. J.,** Transglutaminase and cellular motile events: retardation of proinsulin conversion by glycine methylester, *Bioscience Rep.,* 5, 581, 1985.

44. **Gomis, R., Arbos, M. A., Sener, A., and Malaisse, W. J.,** Glucose-induced activation of transglutaminase in pancreatic islets, *Diabetes Res.,* 3, 115, 1986.

45. **Hellman, B., Idahl, L. Å., Lernmark, Å., Sehlin, J., and Täljedal, I.-B.,** Membrane sulphydryl groups and the pancreatic beta cell recognition of insulin secretagogues, in *Diabetes,* Malaisse, W. J. and Pirart, J., Eds., Excerpta Medica, Amsterdam, 1974, 65.

46. **Henquin, J. C.,** Metabolic control of the potassium permeability in pancreatic islet cells, *Biochem. J.,* 186, 541, 1980.

47. **Cook, D. L., Ikeuchi, M., and Fujimoto, W. Y.,** Lowering of pHi inhibits Ca^{2+}-activated K^+ channels in pancreatic B-cells, *Nature (London),* 311, 269, 1984.

48. **Cook, D. L. and Hales, N.,** Intracellular ATP directly blocks K^+ channels in pancreatic B-cells, *Nature (London),* 311, 271, 1984.

49. **Ashcroft, F. M., Harrison, D. E., and Ashcroft, S. J. H.,** Glucose induces closure of single potassium channels in isolated rat pancreatic β-cells, *Nature (London),* 312, 446, 1984.

50. **Rorsman, P. and Trube, G.,** Glucose dependent K^+ channels in pancreatic β-cells are regulated by intracellular ATP, *Pflugers Arch.,* 405, 305, 1985.

51. **Bellomo, G., Jewell, S. A., Thor, H., and Orrenius, S.,** Regulation of intracellular calcium compartmentation: studies with isolated hepatocytes and t-butyl hydroperoxide, *Proc. Natl. Acad. Sci. U.S.A.,* 79, 6842, 1982.

52. **Panten, U., Zielmann, S., Langer, J., Zünkler, B.-J., and Lenzen, S.**, Regulation of insulin secretion by energy metabolism in pancreatic B-cell mitochondria, *Biochem. J.*, 219, 189, 1984.
53. **Lehninger, A. L., Vercesi, A., and Bababunmi, E. A.**, Regulation of Ca^{2+} release from mitochondria by the oxidation reduction state of pyridine nucleotides, *Proc. Natl. Acad. Sci. U.S.A.*, 75, 1690, 1978.
54. **Valverde, I. and Malaisse, W. J.**, Ionophoretic activity in pancreatic islets, *Biochem. Biophys. Res. Commun.*, 89, 386, 1979.
55. **Dieryck, P., Winand, J., and Malaisse, W. J.**, Ionophoretic activity of lipids extracted from pancreatic islets, *IRCS Med. Sci.*, 9, 190, 1981.
56. **Deleers, M., Mahy, M., and Malaisse, W. J.**, Calcium ionophoresis by pancreatic islet extracts in model membranes, *Int. Biochem.*, 4, 47, 1982.
57. **Anjaneyulu, K., Anjaneyulu, R., and Malaisse, W. J.**, The stimulus-secretion coupling of glucose-induced insulin release. XLIII. Na-Ca countertransport mediated by pancreatic islet native ionophores, *J. Inorg. Biochem.*, 13, 178, 1980.
58. **Valverde, I. and Malaisse, W. J.**, Effects of reduced and oxidized glutathione upon calcium ionophoresis, *IRCS Med. Sci.*, 8, 191, 1980.
59. **Malaisse, W. J., Dufrane, S. P., Mathias, P. C. F., Carpinelli, A. R., Malaisse-Lagae, F., Garcia-Morales, P., Valverde, I., and Sener, A.**, The coupling of metabolic to secretory events in pancreatic islets. The possible role of glutathione reductase, *Biochim. Biophys. Acta*, 844, 256, 1985.
60. **Boyland, E. and Chasseaud, L. F.**, The effect of some carbonyl compounds on rat liver glutathione levels, *Biochem. Pharmacol.*, 19, 1526, 1970.
61. **Wedner, H. J., Simchowitz, L., Stenson, W. F., and Fischman, C. M.**, Inhibition of human polymorphonuclear leukocyte function by 2-cyclohexene-1-one. A role for glutathione in cell activation, *J. Clin. Invest.*, 68, 535, 1981.
62. **Stenson, W. F., Lobos, E., and Wedner, H. J.**, Glutathione depletion inhibits amylase release in guinea pig pancreatic acini, *Am. J. Physiol.*, 244, G273, 1983.
63. **Sener, A., Dufrane, S. P., and Malaisse, W. J.**, The coupling of metabolic to secretory events in pancreatic islets: effects of 2-cyclohexene-1-one upon GSH content and secretory behaviour, *Biochem. Pharmacol.*, 35, 3701, 1986.
64. **Malaisse, W. J., Garcia-Morales, P., Gomis, R., Dufrane, S. P., Mathias, P. C. F., Valverde, I., and Sener, A.**, The coupling of metabolic to secretory events in pancreatic islets: inhibition by 2-cyclohexene-1-one of the secretory response to cyclic AMP and cytochalasin B, *Biochem. Pharmacol.*, 35, 3709, 1986.
65. **Hellerström, C., Andersson, A., and Welsh, M.**, Respiration of the pancreatic B-cell: effects of glucose and 2-aminonorbornane-2-carboxylic acid, *Horm. Metab. Res.*, Suppl. 10, 61, 1980.
66. **Sener, A. and Malaisse, W. J.**, L-leucine and nonmetabolized analogue activate pancreatic islet glutamate dehydrogenase, *Nature (London)*, 288, 187, 1980.
67. **Sener, A., Malaisse-Lagae, F., and Malaisse, W. J.**, Stimulation of pancreatic islet metabolism and insulin release by a nonmetabolizable amino acid, *Proc. Natl. Acad. Sci. U.S.A.*, 78, 5460, 1981.
68. **Malaisse-Lagae, F., Sener, A., Garcia-Morales, P., Valverde, I., and Malaisse, W. J.**, The stimulus-secretion coupling of amino acid-induced insulin release. Influence of a nonmetabolized analogue of leucine on the metabolism of glutamine in pancreatic islets, *J. Biol. Chem.*, 257, 3754, 1982.
69. **Henquin, J. C. and Meissner, H. P.**, Effects of amino acids on membrane potential and $^{86}Rb^+$ fluxes in pancreatic β-cells, *Am. J. Physiol.*, 240, E245, 1981.
70. **Malaisse, W. J. and Sener, A.**, Glucose-induced changes in cytosolic ATP content in pancreatic islets, *Biochim. Biophys. Acta*, 927, 190, 1987.

Energy Requirement of Exocytosis

Chapter 12

CHEMIOSMOTIC EVENTS

Keith W. Brocklehurst and Harvey B. Pollard

TABLE OF CONTENTS

I. INTRODUCTION

The understanding of how metabolic energy is utilized in the process of exocytosis in secretory cells is far from complete. One possible site of action of ATP in secretion became apparent however, with the discovery that the secretory granules of certain tissues possess proton-translocating ATPases. These enzymes use the energy liberated by ATP hydrolysis to pump protons into the interior of secretory granules and thus create an electrochemical proton gradient across the granule membrane. This observation further led to the proposal that the resulting chemiosmotic properties of secretory granules might be involved in the exocytotic mechanism of these tissues.[1]

The "chemiosmotic hypothesis" was originally developed by Mitchell in 1961 to account for the coupling of mitochondrial electron transport to the synthesis of ATP.[2] As electrons are transferred along the respiratory chain in the mitochondrial inner membrane, the energy which is released is utilized to pump protons out of the mitochondrial matrix. The inner mitochondrial membrane is impermeable to protons and so an electrochemical proton gradient is created across the membrane. Protons can however, flow back into the mitochondrion through the ATP synthetase enzyme complex resulting in the synthesis of ATP. The electrochemical proton gradient established across the mitochondrial inner membrane has two components: a pH gradient (Δ pH) with the pH of the matrix being higher than the extramitochondrial environment and a membrane potential ($\Delta\Psi$) with the matrix being negative with respect to the outside. The electrochemical potential ($\Delta\bar{\mu}_{H+}$) or proton-motive force (in mV) is thus given by the following equation:

$$\Delta\bar{\mu}_{H+} = \Delta\psi - \frac{2\cdot3\ RT}{F}\ \Delta pH$$

In the case of secretory granules the direction of the electrochemical proton gradient is opposite to that found with mitochondria because the proton-translocating ATPase present in the granule membrane pumps protons into the intragranular space. The properties of the proton-translocating ATPases of secretory granules and the chemiosmotic properties of the granules resulting from the establishment of an electrochemical proton gradient across the granule membrane will now be discussed in relation to the process of exocytosis.

II. PROTON-TRANSLOCATING Mg^{2+}-DEPENDENT ATPASES OF SECRETORY GRANULES

The discovery of a proton-translocating MgATPase in the secretory granule of adrenal medullary chromaffin cells was made as a result of investigations into the mechanism of catecholamine transport across the granule membrane. While studying catecholamine uptake into isolated chromaffin granules it was found that this process required MgATP.[3,4] It was further shown that catecholamine uptake into granules was associated with ATP hydrolysis and it was proposed that transport was carrier mediated.[5-7] Experiments with chromaffin granule ghosts led to similar conclusions.[8,9] Further evidence confirming a relationship between ATP hydrolysis and catecholamine transport into granules was provided when it was shown that nonhydrolyzable ATP analogs did not support transport.[9-11]

The observation that catecholamine transport in splenic nerve vesicle preparations was inhibited by mitochondrial uncouplers[12] led to insight into the nature of the coupling between ATP hydrolysis and catecholamine transport in chromaffin granules. Catecholamine transport into chromaffin granules was also found to be inhibited by protonophores such as p-trifluoromethoxyphenylhydrazone (FCCP), and the granule membrane MgATPase activity was found to be stimulated by these agents.[13-15] It was also shown that granule MgATPase activity was associated with enhanced fluorescence of the probe 1-anilinonaphthalene-8-sulfonic acid, and this fluorescence enhancement was blocked by mitochondrial uncouplers. By analogy to results from mitochondrial studies it was concluded that the granule MgATPase pumped protons into the granule interior, resulting in a membrane potential which was inside positive. It appeared that the proton gradient thus established was responsible for coupling ATP hydrolysis to catecholamine uptake.

An inwardly directed proton pump would indeed be expected to make the granule transmembrane electrical potential ($\Delta\Psi$) positive inside and to establish a pH gradient (ΔpH) across the granule membrane with the granule interior having a lower pH than the outside. These predictions have indeed proved to be correct. In order to measure the $\Delta\Psi$ of isolated chromaffin granules, the lipid-soluble anion SCN^-,[16-18] the lipid-soluble cation triphenylmethylphosphonium,[19] and a voltage-sensitive dye[20,21] have been employed. In the absence of permeant anions these measurements of $\Delta\Psi$ indicated that the addition of MgATP to granules generated a membrane potential of $+50$ to $+70$ mV inside positive, with an inside negative resting potential at pH 6.9. Measurements of pH in intact granules have been made from the distribution of methylamine or by ^{31}P-NMR (nuclear magnetic resonance) and the results have shown a resting intragranular pH of approximately 5.7 which was relatively independent of the external pH.[16,22-25] A fall of approximately 0.3 pH units over a 30-min period was observed on addition of MgATP in the presence of a permeant anion.[23,24] This pH change was inhibited by uncouplers.[24]

It has proved possible by suitably manipulating the external conditions to induce separate changes in $\Delta\Psi$ and ΔpH. In the presence of a permeant anion, MgATP induced a proton flux into the granules which was electrically neutralized by anion influx. The proton/anion influx resulted in a drop in intragranular pH.[24] However, under these conditions MgATP induced only a small change in membrane potential.[18] Therefore, ΔpH can be changed with only a small change in $\Delta\Psi$. Similarly, when granules were incubated in the absence of a permeant anion, MgATP induced an inside positive membrane potential without a concomitant change in the internal pH.[17,18,24] Presumably, there was no change in intragranular pH under these conditions because of the relatively few positive charges entering before the $\Delta\Psi$ electrically limits further proton entry, and also because of the large buffering capacity of the granule interior. Thus, $\Delta\Psi$ can be altered without changing ΔpH. Chromaffin granule ghosts were found to exhibit similar chemiosmotic properties to intact granules.[26-30]

The nature of the enzyme responsible for pumping protons into chromaffin granules has also been studied in terms of its analogy to the F_0-F_1 ATPase in mitochondria. Under certain conditions the ATP synthetase of the mitochondrial inner membrane can also utilize the energy of ATP hydrolysis to pump protons across the inner membrane, and thus serve as a proton-translocating ATPase, known as the F_0-F_1 ATPase. Although the chromaffin granule proton-translocating MgATPase and the mitochondrial F_0-F_1 ATPase differ in their sensitivities to certain drugs, both enzyme activities can apparently be solubilized by chloroform or dichloromethane, and have been reported to exhibit similar subunit structure and immunologic properties.[31,32] However, it has been suggested that the F_1-type ATPase activity present in chromaffin granule preparations is actually due to mitochondrial contamination and that a separate, anion-sensitive ATPase activity is associated with the granule membrane.[33] This controversy obviously needs to be resolved.

The general properties of chromaffin granules are shared by other types of secretory vesicles. For example, evidence suggesting that the interior of insulin secretory granules of pancreatic B-cells may be more acidic than the cytosol was obtained when it was shown that granule-enriched subcellular fractions from pancreatic islets accumulated the fluorescent dye 9-aminoacridine.[34] Insulin granules in vivo have also been shown to accumulate the base 5-hydroxytryptamine.[35,36] More recently it was shown that insulin granules purified from a rat insulinoma possess an inwardly directed proton-translocating Mg^{2+}-dependent

ATPase.[37] This enzyme was clearly not a mitochondrial contaminant as judged by the distribution of marker proteins on density gradients and the differential sensitivity of the enzyme to certain drugs. The properties of the insulin granule enzyme however were indistinguishable from those of the chromaffin granule enyme and the two types of granules showed similar levels of activity. The resulting chemiosmotic properties of the insulin granule were also very similar to those of the chromaffin granule.[38] The capacity of the insulin granule MgATPase for ATP hydrolysis however far exceeded the energetic requirement for amine transport, as the amine content of insulin granules is much lower than that of chromaffin granules. The authors thus suggested that the insulin granule MgATPase may be involved in secretory granule morphogenesis or exocytosis. The intragranular pH incidentally corresponds to optimal conditions for the crystallization of zinc-insulin hexamers.[38]

Purified neurosecretory vesicles from bovine neurohypophyses also exhibited chemiosmotic properties similar to those of chromaffin granules and insulin granules, consistent with the existence of an inwardly directed proton-translocating MgATPase.[39] The acidic intragranular pH corresponds to the pH of maximal stability of the hormone-neurophysin complex.

The existence of a transmembrane ΔpH in the dense granules of porcine platelets has also been reported,[40] with the inside of the granules being acidic. This observation was later confirmed when it was shown that the dense granules transported 5-hydroxytryptamine in response to the ΔpH across the granule membrane and similar results were obtained with ghosts prepared from the dense granules.[41] It was also shown that a MgATPase activity was present in the ghost membrane and that inhibition of the MgATPase led to a corresponding decrease in 5-hydroxytryptamine transport.

Thus, chromaffin granules, insulin granules, neurohypophyseal secretory vesicles and dense granules of platelets all exhibit very similar chemiosmotic properties. Although the maintenance of a $\Delta\bar{\mu}_{H+}$ across the granule membrane may serve tissue-specific functions such as amine transport in chromaffin and platelet-dense granules and the provision of an optimal chemical environment for the storage of the secretory product in insulin and neurohypophyseal granules, there may be a more general role for chemiosmotic events in the exocytotic process, and this proposal was reinforced by studies of the osmotic lysis of secretory granules.

III. OSMOTIC LYSIS OF SECRETORY GRANULES

In the process of exocytosis, after secretory granules have fused with the plasma membrane of the cell, the barrier separating the granule interior from the extracellular medium undergoes breakage, or fission. This fission step has been modeled in vitro by studying the Cl^-,MgATP-driven osmotic lysis of secretory granules, on the grounds that granule components are involved in both events and both events seem to be associated with a break in the granule membrane.

Studies on the lysis of intact chromaffin granules and on the shrinking or swelling of granule ghosts have been interpreted to indicate a general relative impermeability to cations (e.g. Na^+, K^+, Mg^{2+}, Ca^{2+}, and H^+) and a greater but selective permeability to anions.[42,43] Isolated chromaffin granules incubated under isotonic conditions have been found to undergo MgATP-induced release of their contents in the presence of Cl^- or other permeant anions.[16,44-47] The granule release reaction is due to osmotic lysis as was concluded by earlier workers[48] and more detailed studies showed release could be inhibited by increasing the medium osmotic strength with either salt or sucrose.[16,49] The addition of MgATP to chromaffin granules resulted in the entry of both $[^{35}S]$-SCN^{-} [16] and $^{36}Cl^-$.[50] Cl^-,MgATP-induced granule lysis was also found to be inhibited by mitochondrial uncouplers.[49] It was accordingly proposed that the granule membrane MgATPase catalyzed the inward translocation of protons as counterions to Cl^-, thus raising the osmotic content of the granule and resulting in lysis.[49]

One problem with this proposed mechanism of MgATP-induced granule lysis is that nonhydrolyzable ATP analogs such as App(NH)p caused changes in the transmembrane potential of isolated chromaffin granules[16] and supported granule lysis at 37°C.[10] The V_{max} for release by the analog was however, approximately half that for ATP and the apparent $K_{1/2}$ for App(NH)p was approximately four times greater than that for ATP.[10] Also, ADP was shown to be about half as effective as ATP in inducing both changes in $\Delta\Psi$[16] and in inducing release.[51] Although granule preparations may contain myokinase activity which could explain the ADP effects, ATP-induced changes in $\Delta\Psi$ of granules could be measured at 2°C,[16] a temperature at which enzyme reactions are very slow. ADP did not cause any detectable change in intragranular pH as measured by ^{31}P-NMR.[23] It is thus possible that some granule functions may be regulated by ATP binding rather than hydrolysis. Another explanation for these anomalous results is that only very small amounts of ATP hydrolysis are necessary to generate changes in transmembrane $\Delta\Psi$.

The fact that permeant anion concentration gradients exist across most cell membranes suggests that such gradients would be expected to exist across the membrane barrier separating the granule interior from the extracellular fluid in the fusion state of exocytosis; for example the extracellular Cl^- concentration is approximately 120 mM whereas the intragranular Cl^- concentration has been estimated as being 15 to 40 mM, approximately the same as the cytoplasmic Cl^- concentration.[50] Thus, the anion transport properties of chromaffin granules have been investigated with the expectation that the regulation of anion transport may play a role in secretion.

The first indication that anion entry may be of importance to granule lysis and exocytosis was provided by experiments on granule lysis in the absence of ATP induced by the K^+ ionophore, valinomycin.[52] Thus, in the presence of K^+, valinomycin permitted the electrogenic entry of K^+ into the granule because of the existing low intragranular K^+ concentration.[42] If the K^+ salt of an impermeant anion were used in this system, little K^+ entry took place because of the rapid build up of a K^+ diffusion potential, positive inside. However, if the K^+ salt used were that of a permeant anion, then anion entry accompanied K^+ entry, resulting in the electroneutral cotransport of ions and granule lysis. Based on these experiments an anion permeability series was defined as SCN^-, I^-, $Br^- > Cl^- >$ acetate, F^-, isethionate. This permeability sequence was found to be similar to that obtained for the passive anion permeabilities of chromaffin granule ghosts which was found to be $SCN^- > I^- >$ trichloroacetate $> Br^- > Cl^- > SO_4^{2-} >$ acetate, HCO_3^-, F^-, PO_4^{3-}.[43] Certain anions were also found to support MgATP-induced granule lysis better than others. Thus, SCN^-, I^-, and Br^- were better than Cl^-, while in turn Cl^- was better than acetate, PO_4^{3-}, or isethionate.[6,10,49] It was suggested[49] that MgATP-driven proton translocation was analogous to valinomycin-mediated K^+ movement in inducing net uptake of permeant anions. Anion transport across the granule membrane therefore seemed to be controlled by a selective mechanism, and further evidence of this was provided by observations that the rate of MgATP-induced granule lysis was a saturable function of Cl^- concentration and that isethionate, an impermeant anion, acted as a competitive inhibitor of lysis with respect to Cl^-.[50] These authors thus proposed a mechanism for exocytosis based on ATP-driven anion entry into granules fused with the plasma membrane after stimulation of the chromaffin cell with secretagogue.

A biophysical model has been constructed to describe the behavior of isolated chromaffin granules on undergoing Cl^-,MgATP-induced lysis.[53] The model relies on the assumption that the granule release reaction represents osmotic lysis due to the MgATPase-dependent influx of protons and osmotically active Cl^- ions. The consequences of this influx were predicted from osmotic fragility curves determined by suspending granules in hypotonic media. Turbidity measurements of granule suspensions were used to fit the model parameters. The model was found to successfully describe the time course, Cl^--dependence, ATP-dependence, and osmotic strength suppression of the release reaction as measured either by turbidity changes or epinephrine release.

In the erythrocyte membrane, anion transport in the form of anion exchange is mediated by a specific membrane protein (band III) and this transport system is competitively inhibited by several drugs including 4-acetamido-4'-isothiocyanostilbene 2,2'-disulfonic acid (SITS), probenecid, and pyridoxal phosphate,[54] these drugs being relatively impermeant aromatic anions. It was of particular interest to discover that these compounds also blocked MgATP-induced granule lysis observed in the presence of Cl^-, and this inhibition was competitive with respect to Cl^-, analogous to the competitive effect of isethionate on Cl^--supported release.[50] The K_i values of SITS, probenecid, and pyridoxal phosphate with respect to Cl^- for granule lysis were 40 μM, 125 μM, and 3.6 mM respectively, values similar to those found in the erythrocyte system. SITS was also shown to inhibit MgATP-induced uptake of $^{36}Cl^-$ into granules. It was thus concluded that the mechanism of Cl^-,MgATP-induced lysis of chromaffin granules probably included anion entry through a site pharmacologically analogous to band III in erythrocytes.

The anion transport blockers have since been used to characterize various mechanisms of anion transport in chromaffin granules, and three groups of anions relating to these mechanisms have been distinguished.[55,56] The classification of different anions was based on their ability to support granule lysis under different conditions. Class I consists of Cl^-, Br^-, and I^-, all relatively permeant anions; class II includes PO_4^{3-} and isethionate which are impermeant anions; and class III includes F^-, SO_4^{2-}, and acetate, the permeabilities of which are dependent upon the conditions used. The ionophore nigericin mediates granule lysis in the presence of K^+ by allowing the exchange of intragranular osmotically inactive protons for medium K^+ whereas NH_4^+ mediates lysis by permeating the granule membrane as NH_3 and subsequently becoming protonated and trapped within the granule. Both these lytic mechanisms were found to be independent of anion permeation as all three anion classes were permissive and lysis was not inhibited by the anion transport blocking drugs.

Both valinomycin/K^+-induced lysis and MgATP-induced lysis were dependent on the presence of the class I permeant anions, but the former mechanism in contrast to the latter was not inhibited by the anion transport blockers. Although class II impermeant anions could inhibit MgATP-induced lysis in the presence of class I anions this was not found to be the case for valinomycin/K^+-induced release. Thus, there appeared to be two distinct types of anion transport site with similar anion selectivity but with differential inhibitor sensitivity.

Granule lysis could also be induced by imposing an artificially induced inwardly directed proton gradient across the granule membrane, this being achieved by incubating granules in low pH medium (pH < 5.5). As with MgATP-induced lysis, class I anions stimulated lysis in a blocker-sensitive manner, and class II anions were ineffective in supporting lysis but could inhibit lysis in the presence of class I anions. Thus, anion-permeation induced by low external pH and by MgATP appear to involve the same anion transport site.

Class III anions were also found to be effective at inducing lysis at low external pH and this effect was not blocker sensitive. It is possible that this class of anions permeates the granule membrane as uncharged protonated species and becomes deprotonated and trapped within the granule. A possible analogous situation involves the permeant anion SCN^- which can support granule lysis whenever anion permeability is required as in lysis induced by valinomycin/K^+, MgATP, or low external pH. However SCN^--supported lysis with MgATP or at low pH is not blocker sensitive and so probably the lipophilic nature of SCN^- allows it to permeate the granule membrane without the need of a specific transport site.

Thus, these studies led the authors to conclude that only Cl^-, Br^-, and I^- use the blocker-sensitive anion transport site and that this site is only used in conjunction with proton entry driven either by MgATP hydrolysis or an artificially imposed proton gradient across the granule membrane. It is possible that the proton transport site and the blocker-sensitive anion transport site are intimately related and possibly part of a single macromolecular complex that includes the MgATPase. This idea is supported by the finding that the granule MgATPase activity, even in detergent-solubilized form, is potentiated by permeant but not impermeant anions, and is inhibited by SITS and pyridoxal phosphate.[55]

IV. STUDIES ON THE ROLE OF CHEMIOSMOTIC EVENTS IN EXOCYTOSIS USING INTACT CELLS

If the chromaffin granule lysis reaction is representative of the fission step in exocytosis then it should be possible to make certain predictions about the exocytotic process based on the experimental results obtained with isolated granules. In the chemiosmotic model of secretion[1] it was thus proposed that when secretory granules form a fusion complex with the plasma membrane of the cell an anion transport site derived from the granule membrane might span the membrane complex now separating the intragranular space from the extracellular medium. The resulting transport of anions down their concentration gradient into the granule coupled to the proton-translocating MgATPase in the granule membrane might then provoke fission of the fused membrane complex by osmotic lysis. It should be noted that in an artificial system, the fusion of liposomes with planar lipid bilayers requires an osmotic gradient across the vesicle membrane to allow entry of H_2O and vesicle swelling.[57,58] Thus, in intact cells for example it might be expected that exocytosis would be inhibited by (1) the anion transport-blocking drugs, (2) by the absence of specific permeant anions from the bathing medium, (3) by hypertonic media, and (4) by protonophores.

These predictions were first tested on human platelets.[59] The secretion of 5-hydroxytryptamine from human platelets stimulated by thrombin or the Ca^{2+}-ionophore, A23187, was found to be inhibited by the anion transport blockers SITS, pyridoxal phosphate, probenecid, and suramin. However, replacement of NaCl in the medium by either sucrose or sodium isethionate did not inhibit secretion. Instead a reduction in the medium pH was found to inhibit platelet secretion and inhibition by the anion transport blockers was found to be competitive with respect to OH^-. The K_i values for SITS, probenecid, pyridoxal phosphate, and suramin were 28, 335, 56, and 0.9 μM respectively with respect to OH^-. Thus, the activity series for inhibition of platelet secretion was suramin $>>$ SITS, pyridoxal phosphate $>>$ probenecid with isethionate being inactive, whereas for inhibition of chromaffin granule lysis the series was SITS $>$ probenecid $>$ suramin $>$ pyridoxal phosphate $>>$ isethionate. It is possible that the differences in activity of the drugs in the two systems relate to anion specificity. 5-Hydroxytryptamine release from platelets was also suppressed by increasing extracellular osmotic strength, and the relationship between suppression and osmotic strength was quantitatively similar to that observed for chromaffin granule lysis. The protonophore FCCP was also found to inhibit platelet secretion. It was thus concluded[59] that platelets and chromaffin granules were similar in terms of the osmotic basis of release in these systems, although they appeared to differ in the specific anions involved.

The chemiosmotic hypothesis of secretion was also investigated in dissociated parathyroid cells[60] from which parathyroid hormone (PTH) secretion can be elicited by exposure to a low extracellular Ca^{2+} concentration (0.5 mM).[61] SITS and probenecid were found to inhibit PTH secretion almost completely whereas replacement of NaCl by either sucrose or sodium isethionate caused a 70% inhibition of secretion. The inhibition of secretion by SITS and probenecid was competitive with respect to Cl^-, and each drug had a K_i between 400 to 600 μM. Raising the osmotic strength of the medium completely inhibited secretion. Various cation replacements for Na^+ had no effect on PTH release but FCCP completely blocked secretion. The observation that anion transport blockers inhibited PTH secretion almost completely whereas the omission of Cl^- or its replacement with isethionate only inhibited release by 70% led to the interpretation that an alternative anion to Cl^- might be present in the medium. Lowering the medium pH was found to inhibit PTH secretion and probenecid was also found to be a competitive inhibitor of release with respect to OH^-. Thus, it was concluded that OH^- might also be a permeant anion for this system.

Obviously the secretory cell of most relevance for comparing the properties of chromaffin granule lysis with those of secretion from intact cells, is the chromaffin cell, and this cell has provided both evidence in favor and against the chemiosmotic model of secretion. Using bovine chromaffin cells isolated by collagenase digestion of adrenal medullae, it was shown that increasing the osmotic strength of the medium with either NaCl or sucrose resulted in suppression of veratridine- or acetylcholine-induced epinephrine secretion in an almost quantitatively identical manner to the osmotic suppression of Cl^-,MgATP-induced lysis of isolated chromaffin granules.[62] Epinephrine secretion occurred only in the presence of a permeant anion such as Cl^-, and veratridine-induced secretion was inhibited when medium NaCl was replaced by sodium isethionate. Using different monovalent anions, veratridine was found to support epinephrine release according to the activity series Br^-, I^-, $NO_3^- >$ methylsulfate, $SCN^- > Cl^- >$ acetate $>>$ isethionate. A similar series, except for the potency of NO_3^- was seen when A23187 was the secretagogue. However, although the anion series for MgATP-dependent granule lysis was also qualitatively similar, there was a poor quantitative correlation between the anion dependence of chemiosmotic granule lysis and secretion from intact cells.

FCCP was found to inhibit veratridine-induced secretion from chromaffin cells in a concentration range previously shown to block Cl^-,MgATP-driven granule lysis. However, other agents which act on mitochondrial energy metabolism, such as CN^-, azide, and rotenone, also inhibited secretion whereas they were without effect on Cl^-,MgATP-driven granule lysis. Therefore, FCCP inhibition of epinephrine secretion could not be attributed unequivocally in terms of its effect on granules.

Specific differences were also noted between the effects of the anion channel blocking drugs on chromaffin granule lysis in vitro and on secretion from intact cells. Probenecid and pyridoxal phosphate inhibited veratridine-induced secretion from chromaffin cells, although only at relatively high concentrations, while in contrast to their effect on the granule lysis reaction, SITS and other stilbene disulfonates were inactive. Therefore, while the properties of secretion from chromaffin cells resembled some of the properties of Cl^-,MgATP-induced epinephrine release from isolated secretory granules, a number of differences were found. Thus, the authors concluded that while some mechanistic relationships may exist between the exocytotic release event in chromaffin cells and the chemiosmotic lysis reaction of isolated chromaffin granules, an understanding of the energetics of exocytosis awaits the discovery of the reasons for the quantitative differences between the two systems.

The effects of changes in osmolality on the stability and function of cultured bovine chromaffin cells have also been extensively investigated.[63] Measurements of cell volume in media of differing osmolality showed that the cells behaved as osmometers with the intracellular osmolality equilibrating rapidly with the osmolality of the bathing medium. Hyperosmotic solutions inhibited both nicotinic agonist-stimulated and K^+-stimulated catecholamine secretion but were without effect on nicotinic agonist-induced Ca^{2+} uptake and only weakly inhibited K^+-induced Ca^{2+} uptake. Thus, increased osmolality inhibited both nicotinic agonist- and K^+-induced secretion at a step distal to Ca^{2+} entry into the cell, and it was concluded that osmotic forces acting subsequent to Ca^{2+} entry may play a role in secretion. This study also demonstrated the fact that the osmotic fragility of chromaffin granules in vitro is greater than that of granules *in situ*. Reducing the osmolality of the medium in which isolated chromaffin granules were suspended from 310 to 210 mOsm caused the release of more than 75% of their catecholamine. By contrast, reduction of the osmolality to below 165 mOsm was necessary before evidence of substantial intracellular granular lysis was apparent. This disquieting observation means that what is true of isolated granules may not necessarily be true of granules in their intracellular environment.

An extensive investigation into the possible involvement in exocytosis of the proton electrochemical gradient present across the chromaffin granule membrane has also been made using cultured bovine chromaffin cells.[64] The authors adopted several experimental approaches in order to reduce the proton electrochemical gradient of intracellular granules. Firstly, ammonium or methylamine ions, or nigericin were found to reduce the pH gradient across the granule membrane as judged by cell/medium methylamine concentration ratios or by ^{31}P-NMR of intragranular ATP. An encouraging observation was that the pH inside chromaffin granules within cells was found to be 5.3 under resting conditions. However, these treatments were without effect on veratridine-induced catecholamine secretion and nigericin did not alter secretion induced by nicotinic receptor stimulation. Secondly, FCCP at a concentration of 1 μM, which uncoupled mitochondria within chromaffin cells as judged by O_2 consumption measurements, was found to have no inhibitory effect on carbachol-induced secretion. Finally, dicyclohexylcarbodiimide (DCCD) was found to have no effect on K^+-induced catecholamine release under conditions which blocked coupled transport of [3H]-norepinephrine into chromaffin granules isolated from cells that had been previously incubated with DCCD. DCCD is an irreversible inhibitor of proton-translocating ATPases and has been shown to inhibit the chromaffin granule membrane MgATPase, the MgATP-induced increase in granule membrane potential, and MgATP-induced catecholamine uptake in isolated chromaffin granules.[15] The authors concluded that these results make unlikely the possibility that the combination of the proton-translocating MgATPase, the granule membrane electrical potential and extracellular anions is involved in the mechanism of exocytosis. Although these data are fairly convincing, if one were to play devil's advocate there are several possible criticisms of these experiments. First, ammonium, methylamine and nigericin reduce the ΔpH across the chromaffin granule membrane but not $\Delta\psi$. Second, there was no demonstrated effect of FCCP on intracellular granules; but as FCCP does not alter the internal pH of isolated granules, then a pH gradient may still exist between the granules and the cytosol or extracellular medium. Third, there was also no demonstrated effect of DCCD on the ΔpH across the granule membrane and as mentioned above the efficacy of this agent in the cells was judged by its effect on coupled [3H]-norepinephrine uptake into granules isolated from DCCD-treated cells. However, part of the inhibition of MgATP-induced catecholamine uptake in isolated granules by DCCD may be due to a direct effect on the catecholamine transporter.[65] Also the authors point out that conditions could be found where methylamine, nigericin, and FCCP all inhibited catecholamine secretion: high methylamine concentrations inhibited secretion but also reduced the cytosolic ATP concentration; increasing the nigericin concentration from 1 to 10 μM caused partial inhibition of secretion but this was not correlated with a substantial change in the intragranular pH; and FCCP concentrations greater than 1 μM inhibited secretion but were larger than necessary to maximally uncouple intracellular mitochondria. Therefore, inhibition of secretion under these conditions was probably related to effects other than on the chromaffin granule proton electrochemical potential. This set of experiments therefore although not completely killing the chemiosmotic hypothesis as applied to the chromaffin cell has left it in critical condition.

The chemiosmotic hypothesis as applied to secretion from intact cells has also been investigated in several other cell types. Lysosomal enzyme release from human neutrophils stimulated with immune complexes was inhibited by SITS, H,H'-diisothiocyano-2,2'-stilbene disulfonic acid (DIDS), and pyridoxal phosphate.[66] A23187-induced release was also inhibited by SITS and DIDS. Neither the permeant anion(s) nor the role of anion influx in degranulation was identified however, although influx of Cl^-, OH^-, or PO_4^{3-} did not seem to be important. Permeant anions supported antigen-induced histamine release from human basophils with the following activity series: acetate $> Br^-$, $I^- > Cl^-$.[67] Isethionate and SO_4^{2-} did not support histamine release. SITS and probenecid did not inhibit IgE-mediated histamine release and isoosmotic solutions of several sugars were capable of supporting antigen- or anti-IgE-induced histamine release in the presence of Ca^{2+}. Increasing the osmolarity of the medium by adding more NaCl/KCl enhanced antigen- or anti-IgE-induced histamine release. These results therefore do not support the hypothesis that exocytosis from basophils is dependent on anions or that it is the result of osmotic lysis.

Finally the predictions of the chemiosmotic hypothesis of secretion have been applied to insulin secretion from pancreatic islets. Insulin release induced by glucose or α-ketoisocaproate from isolated rat pancreatic islets was inhibited when extracellular Cl^- was replaced with isethionate or SO_4^{2-}, when the extracellular osmotic strength was increased by addition of sucrose, and when the islets were exposed to probenecid or DIDS.[68-70] The inhibition of glucose-induced insulin release by Cl^- substitution was associated with a modest decrease in glucose oxidation but no significant change in glucose-stimulated $^{45}Ca^{2+}$ net uptake by the islets.[69] In the isolated perfused rat pancreas the isethionate- or sucrose-induced inhibition of glucose-stimulated insulin release was a rapid and rapidly reversible phenomenon.[69] However, Cl^- substitution by isethionate in this system inhibited the second phase of the secretory response to glucose greater than the first phase, and failed to inhibit the insulin response to gliclazide. Similarly, using perifused rat islets, Cl^- substitution by isethionate failed to inhibit the first phase of glucose-induced insulin release or tolbutamide-stimulated secretion.[71] Thus, although the chemiosmotic hypothesis for exocytosis might be applicable to the process of insulin secretion, further evidence is required to fully substantiate such a mechanism.

V. STUDIES ON THE ROLE OF CHEMIOSMOTIC EVENTS IN EXOCYTOSIS USING PERMEABILIZED CELLS

In an attempt to gain direct access to the secretory machinery of the cell and thus bypass the permeability barrier imposed by the plasma membrane, chromaffin cells have been permeabilized by subjecting them to high-voltage discharge. Catecholamine release from these cells can be induced by low Ca^{2+} concentrations (in the low micromolar range) and requires MgATP.[72-75] The standard medium used to study catecholamine release from these cells contains glutamate as the major anion and when glutamate was replaced by Cl^-, Ca^{2+}-dependent release was inhibited. The effectiveness of various anions at inhibiting Ca^{2+}-dependent release was in the order $SCN^- > Br^- > Cl^- >$ acetate $>$ glutamate which follows the order of the lyotropic series. It was thus suggested that the inhibitory anions may bind to and disrupt some part of the release machinery, and that this release machinery might be involved in regulating exocytosis in vivo. There was no effect of SITS or DIDS at a concentration of 0.1 mM on Ca^{2+}-evoked release. Glutamate has been shown to support catecholamine release from intact chromaffin cells approximately 75% as well as Cl^- when sodium glutamate was substituted for NaCl, and glutamate was also able to support MgATP-dependent chromaffin granule lysis, though substantially less well than Cl^-.[62] Ca^{2+} activation of release from permeabilized cells was also found to be essentially normal in buffered isotonic sucrose but increasing the osmotic pressure of the medium by raising the sucrose concentration caused inhibition of Ca^{2+}-dependent release and may reflect the involvement of an osmotically active step in the secretory process.

The average internal pH of chromaffin granules in high-voltage permeabilized cells was estimated to be 5.8 from measurements of [^{14}C]-methylamine uptake into the cells.[76] Methylamine accumulation was greatly reduced by the presence of ammonium ions or monensin (a proton/cation exchanger) but unaffected by the proton pump blocker trimethyltin or FCCP. [^{14}C]-SCN accumulation measurements were also made and used as a measurement of the granule membrane potential. The mean membrane potential was reduced in the absence of added ATP and in the presence of trimethyltin or FCCP. Neither the intragranular pH nor the granule membrane potential was affected by Ca^{2+} over the range of free Ca^{2+} concentrations that activate release, except when large amounts of catecholamine were released. In these cases both the methylamine and SCN$^-$ spaces were reduced, this being consistent with these compounds being accumulated in secretory granules. Exposure of leaky cells to ammonium concentrations that strongly alkalinize the granule core failed to block Ca^{2+}-dependent release, although there was a small reduction in Ca^{2+}-dependent release at ammonium concentrations that greatly reduce the pH gradient between the vesicle core and the cytosol.

Both Ca^{2+}-dependent release and [^{14}C]-SCN$^-$ accumulation were dependent on MgATP but the latter was sensitive to lower MgATP concentrations than the former. There were also differences in the nucleotide specificity of the two processes, Ca^{2+}-dependent release being very specific for ATP, whereas ATP, GTP, UTP, and ITP were almost equally as effective at activating [^{14}C]-SCN$^-$ accumulation. Ca^{2+}-dependent release persisted when the granule proton pump and SCN$^-$ accumulation were inhibited by trimethyltin, although high trimethyltin concentrations did cause a small reduction in Ca^{2+}-dependent release. Ca^{2+}-dependent release also persisted in the presence of FCCP concentrations that collapsed the granule membrane potential, but very high FCCP concentrations also caused a small inhibition of Ca^{2+}-dependent release. There were no significant alterations in the Ca^{2+} activation curve of release under conditions where the vesicle pH gradient or membrane potential were largely collapsed. Ca^{2+}-dependent release also took place when both the granule pH gradient was collapsed with ammonium and the granule membrane potential was collapsed with trimethyltin or FCCP.

Therefore, Ca^{2+}-dependent release from high-voltage permeabilized cells was little affected by the magnitudes of the secretory vesicle pH gradient and membrane potential. However, as the authors point out this conclusion is heavily dependent on the interpretation of [^{14}C]-SCN$^-$ and [^{14}C]-methylamine spaces in terms of the membrane potential and internal pH of secretory vesicles respectively. Although contributions to these spaces from other compartments such as lysosomes cannot be excluded, it is likely that chromaffin granules contribute a significant proportion of these spaces. As Ca^{2+}-dependent release persists with both methylamine and SCN$^-$ spaces close to the [^{3}H]-H$_2$O space, it would seem that the chromaffin granule pH gradient and membrane potential are not essential for this process. However, the reduction in Ca^{2+}-dependent release seen at high concentrations of the various agents used to collapse the pH gradient and membrane potential across the granule membrane may suggest a small modulatory role for these in exocytosis. It must be remembered however, that in permeabilized cells there is no membrane potential or ion gradients across the plasma membrane and so the release event in these cells may not be truly representative of that which occurs in intact cells.

ATP has also been found to be required for Ca^{2+}-dependent catecholamine secretion from digitonin-permeabilized chromaffin cells[77,78] and Ca^{2+}-dependent insulin secretion from permeabilized pancreatic islets,[79] but as with the case of the high-voltage permeabilized chromaffin cells the requirement for ATP in these systems is not understood.

VI. CONCLUSIONS

The chemiosmotic hypothesis of secretion was formulated as a result of the chemiosmotic properties exhibited by isolated chromaffin granules. Thus, the observation that chromaffin granules undergo osmotic lysis in the presence of Cl$^-$ and an electrochemical proton gradient across the granule membrane established by the proton-translocating MgATPase, formed the experimental basis for several predictions as to the behavior of secretory cells. However, it became apparent with the results subsequently obtained with intact and permeabilized cells that there were large differences between the predicted and observed secretory properties of these cells.

An interesting finding to emerge from these studies however was that isolated chromaffin granules appear to display different properties to granules present either in intact cells or even permeabilized cells. Thus, chromaffin granules within intact chromaffin cells appear to be more resistant to osmotic lysis than isolated granules,[63] and chromaffin granules present within electrically permeabilized cells do not appear to undergo Cl^-,MgATP-induced osmotic lysis.[73-75] It may be that granules undergo a "preparation catastrophe" during their isolation which causes them to express characteristics only normally revealed under stress, as may indeed occur at the cell surface when the granule membrane fuses with the plasma membrane to form the fusion complex. Further investigation is required to try and elucidate the reasons why isolated granules show different properties from granules in their natural environment.

Thus, although the chemiosmotic hypothesis as originally formulated seems not to explain in quantitative detail the properties of secretory cells, it has proved useful in providing a stimulus to research into the energetics of secretion, and it is equally evident that there is no alternative hypothesis presently available to explain how exocytosis is powered.

REFERENCES

1. **Pollard, H. B., Pazoles, C. J., Creutz, C. E., and Zinder, O.,** The chromaffin granule and possible mechanisms of exocytosis, *Int. Rev. Cytol.*, 58, 159, 1979.
2. **Mitchell, P.,** Coupling of phosphorylation to electron and hydrogen transfer by a chemi-osmotic type of mechanism, *Nature (London)*, 191, 144, 1961.
3. **Kirshner, N.,** Uptake of catecholamines by a particulate fraction of the adrenal medulla, *J. Biol. Chem.*, 237, 2311, 1962.
4. **Carlsson, A., Hillarp, N. A., and Waldeck, B.,** Analysis of the Mg^{++}-ATP dependent storage mechanism in the amine granules of the adrenal medulla, *Acta Physiol. Scand.*, Suppl. 215, 1, 1963.
5. **Taugner, G.,** The membrane of catecholamine storage vesicles of adrenal medulla. Catecholamine fluxes and ATPase activity, *Naunyn-Schmiedeberg's Arch. Pharmakol.*, 270, 392, 1971.
6. **Taugner, G.,** The effects of univalent anions on catecholamine fluxes and adenosine triphosphatase activity in storage vesicles from the adrenal medulla, *Biochem. J.*, 130, 969, 1972.
7. **Slotkin, T. A.,** Hypothetical model of catecholamine uptake into adrenal medullary storage vesicles, *Life Sci.*, 13, 675, 1973.
8. **Phillips, J. H.,** Transport of catecholamines by resealed chromaffin-granule 'ghosts', *Biochem. J.*, 144, 311, 1974.
9. **Phillips, J. H.,** Steady-state kinetics of catecholamine transport by chromaffin-granule 'ghosts', *Biochem. J.*, 144, 319, 1974.
10. **Hoffman, P. G., Zinder, O., Bonner, W. M., and Pollard, H. B.,** Role of ATP and β-γ-iminoadenosinetriphosphate in the stimulation of epinephrine and protein release from isolated adrenal secretory vesicles, *Arch. Biochem. Biophys.*, 176, 375, 1976.
11. **Hoffman, P. G., Zinder, O., Nikodijevik, O., and Pollard, H. B.,** ATP-stimulated transmitter release and cyclic AMP synthesis in isolated chromaffin granules, *J. Supramol. Struct.*, 4, 181, 1976.
12. **von Euler, U. S. and Lishajko, F.,** Effects of some metabolic co-factors and inhibitors on transmitter release and uptake in isolated adrenergic nerve granules, *Acta Physiol. Scand.*, 77, 298, 1969.
13. **Bashford, C. L., Radda, G. K., and Ritchie, G. A.,** Energy-linked activities of the chromaffin granule membrane, *FEBS Lett.*, 50, 21, 1975.
14. **Bashford, C. L., Casey, R. P., Radda, G. K., and Ritchie, G. A.,** The effect of uncouplers on catecholamine incorporation by vesicles of chromaffin granules, *Biochem. J.*, 148, 153, 1975.
15. **Bashford, C. L., Casey, R. P., Radda, G. K., and Ritchie, G. A.,** Energy-coupling in adrenal chromaffin granules, *Neuroscience*, 1, 399, 1976.
16. **Pollard, H. B., Zinder, O., Hoffman, P. G., and Nikodijevik, O.,** Regulation of the transmembrane potential of isolated chromaffin granules by ATP, ATP analogs, and external pH, *J. Biol. Chem.*, 251, 4544, 1976.
17. **Holz, R. W.,** Evidence that catecholamine transport into chromaffin vesicles is coupled to vesicle membrane potential, *Proc. Natl. Acad. Sci. U.S.A.*, 75, 5190, 1978.

18. **Johnson, R. G. and Scarpa, A.,** Protonmotive force and catecholamine transport in isolated chromaffin granules, *J. Biol. Chem.,* 254, 3750, 1979.
19. **Holz, R. W.,** Measurement of membrane potential of chromaffin granules by the accumulation of triphenylmethylphosphonium cation, *J. Biol. Chem.,* 254, 6703, 1979.
20. **Ogawa, M. and Inouye, A.,** Responses of the transmembrane potential coupled to the ATP-evoked catecholamine release in isolated chromaffin granules, *Jpn. J. Physiol.,* 29, 309, 1979.
21. **Salama, G., Johnson, R. G., and Scarpa, A.,** Spectrophotometric measurements of transmembrane potential and pH gradients in chromaffin granules, *J. Gen. Physiol.,* 75, 109, 1980.
22. **Johnson, R. G. and Scarpa, A.,** Internal pH of isolated chromaffin vesicles, *J. Biol. Chem.,* 251, 2189, 1976.
23. **Pollard, H. B., Shindo, H., Creutz, C. E., Pazoles, C. J., and Cohen, J. S.,** Internal pH and state of ATP in adrenergic chromaffin granules determined by ^{31}P nuclear magnetic resonance spectroscopy, *J. Biol. Chem.,* 254, 1170, 1979.
24. **Casey, R. P., Njus, D., Radda, G. K., and Sehr, P. A.,** Active proton uptake by chromaffin granules: observation by amine distribution and phosphorus-31 nuclear magnetic resonance techniques, *Biochemistry,* 16, 972, 1977.
25. **Njus, D., Sehr, P. A., Radda, G. K., Ritchie, G. A., and Seeley, P. J.,** Phosphorus-31 nuclear magnetic resonance studies of active proton translocation in chromaffin granules, *Biochemistry,* 17, 4337, 1978.
26. **Phillips, J. H. and Allison, Y. P.,** Proton translocation by the bovine chromaffin-granule membrane, *Biochem. J.,* 170, 661, 1978.
27. **Johnson, R. G., Pfister, D., Carty, S. E., and Scarpa, A.,** Biological amine transport in chromaffin ghosts. Coupling to the transmembrane proton and potential gradients, *J. Biol. Chem.,* 254, 10963, 1979.
28. **Johnson, R. G., Carty, S. E., and Scarpa, A.,** Proton: substrate stoichiometries during active transport of biogenic amines in chromaffin ghosts, *J. Biol. Chem.,* 256, 5773, 1981.
29. **Knoth, J., Handloser, K., and Njus, D.,** Electrogenic epinephrine transport in chromaffin granule ghosts, *Biochemistry,* 19, 2938, 1980.
30. **Apps, D. K., Pryde, J. G., and Phillips, J. H.,** Both the transmembrane pH gradient and the membrane potential are important in the accumulation of amines by resealed chromaffin-granule 'ghosts', *FEBS Lett.,* 111, 386, 1980.
31. **Apps, D. K. and Schatz, G.,** An adenosine triphosphatase isolated from chromaffin-granule membranes is closely similar to F_1-adenosine triphosphatase of mitochondria, *Eur. J. Biochem.,* 100, 411, 1979.
32. **Apps, D. K.,** Proton-translocating ATPase of chromaffin granule membranes, *Fed. Proc.,* 41, 2775, 1982.
33. **Cidon, S. and Nelson, N.,** A novel ATPase in the chromaffin granule membrane, *J. Biol. Chem.,* 258, 2892, 1983.
34. **Abrahamsson, H. and Gylfe, E.,** Demonstration of a proton gradient across the insulin granule membrane, *Acta Physiol. Scand.,* 109, 113, 1980.
35. **Ekholm, R., Ericson, L. E., and Lundquist, I.,** Monoamines in the pancreatic islets of the mouse. Subcellular localization of 5-hydroxytryptamine by electron microscopic autoradiography, *Diabetologia,* 7, 339, 1971.
36. **Hellman, B., Lernmark, A., Sehlin, J., and Taljedal, I.-B.,** Transport and storage of 5-hydroxytryptamine in pancreatic B-cells, *Biochem. Pharmacol.,* 21, 695, 1972.
37. **Hutton, J. C. and Peshavaria, M.,** Proton-translocating Mg^{2+}-dependent ATPase activity in insulin-secretory granules, *Biochem. J.,* 204, 161, 1982.
38. **Hutton, J. C.,** The internal pH and membrane potential of the insulin-secretory granule, *Biochem. J.,* 204, 171, 1982.
39. **Russel, J. T. and Holz, R. W.,** Measurement of ΔpH and membrane potential in isolated neurosecretory vesicles from bovine neurohypophyses, *J. Biol. Chem.,* 256, 5950, 1981.
40. **Johnson, R. G., Scarpa, A., and Salganicoff, L.,** The internal pH of isolated serotonin containing granules of pig platelets, *J. Biol. Chem.,* 253, 7061, 1978.
41. **Wilkins, J. A. and Salganicoff, L.,** Participation of a transmembrane proton gradient in 5-hydroxytryptamine transport by platelet dense granules and dense-granule ghosts, *Biochem. J.,* 198, 113, 1981.
42. **Johnson, R. G. and Scarpa, A.,** Ion permeability of isolated chromaffin granules, *J. Gen. Physiol.,* 68, 601, 1976.
43. **Phillips, J. H.,** Passive ion permeability of the chromaffin-granule membrane, *Biochem. J.,* 168, 289, 1977.
44. **Oka, M., Ohuchi, T., Yoshida, H., and Imaizumi, R.,** Effect of adenosine triphosphate and magnesium on the release of catecholamines from adrenal medullary granules, *Biochim. Biophys. Acta,* 97, 170, 1965.
45. **Poisner, A. M. and Trifaro, J. M.,** The role of ATP and ATPase in the release of catecholamines from the adrenal medulla. I. ATP-evoked release of catecholamines, ATP, and protein from isolated chromaffin granules, *Mol. Pharmacol.,* 3, 561, 1967.
46. **Lishajko, F.,** Influence of chloride ions and ATP-Mg^{++} on the release of catecholamines from isolated adrenal medullary granules, *Acta Physiol. Scand.,* 75, 255, 1969.

47. **Pollard, H. B., Zinder, O., and Hoffman, P. G.,** Occurrence and properties of chromaffin granules, in *Biological Handbook, I: Cell Biology,* Altman, P. L. and Katz, D. D., Eds., Fed. Am. Soc. Exp. Biol., Bethesda, Md., 1976, 358.

48. **Viveros, O. H.,** Mechanism of secretion of catecholamines from adrenal medulla, in *Handbook of Physiology,* Section 7, Vol. 6, Am. Physiol. Soc., Washington D.C., 1975, 389.

49. **Casey, R. P., Njus, D., Radda, G. K., and Sehr, P. A.,** Adenosine triphosphate-evoked catecholamine release in chromaffin granules. Osmotic lysis as a consequence of proton translocation, *Biochem. J.,* 158, 583, 1976.

50. **Pazoles, C. J. and Pollard, H. B.,** Evidence for stimulation of anion transport in ATP-evoked transmitter release from isolated secretory vesicles, *J. Biol. Chem.,* 253, 3962, 1978.

51. **Izumi, F., Kashimoto, T., Miyashita, T., Wada, A., and Oka, M.,** Involvement of membrane associated protein in ADP-induced lysis of chromaffin granules, *FEBS Lett.,* 78, 177, 1977.

52. **Dolais-Kitabgi, J. and Perlman, R. L.,** The stimulation of catecholamine release from chromaffin granules by valinomycin, *Mol. Pharmacol.,* 11, 745, 1975.

53. **Creutz, C. E. and Pollard, H. B.,** A biophysical model of the chromaffin granule. Accurate description of the kinetics of ATP and Cl^- dependent granule lysis, *Biophys. J.,* 31, 255, 1980.

54. **Cabantchik, Z. I., Knauf, P. A., and Rothstein, A.,** The anion transport system of the red blood cell. The role of membrane protein evaluated by the use of 'probes', *Biochim. Biophys. Acta,* 515, 239, 1978.

55. **Pazoles, C. J., Creutz, C. E., Ramu, A., and Pollard, H. B.,** Permeant anion activation of MgATPase activity in chromaffin granules. Evidence for direct coupling of proton and anion transport, *J. Biol. Chem.,* 255, 7863, 1980.

56. **Pazoles, C. J.,** Anion and proton transport in chromaffin granules, *Fed. Proc.,* 41, 2769, 1982.

57. **Zimmerberg, J., Cohen F. S., and Finkelstein, A.,** Fusion of phospholipid vesicles with planar phospholipid bilayer membranes. I. Discharge of vesicle contents across the planar membrane, *J. Gen. Physiol.,* 75, 241, 1980.

58. **Cohen, F. S., Zimmerberg, J., and Finkelstein, A.,** Fusion of phospholipid vesicles with planar phospholipid bilayer membranes. II. Incorporation of a vesicle membrane marker into the planar membrane, *J. Gen. Physiol.,* 75, 251, 1980.

59. **Pollard, H. B., Tack-Goldman, K., Pazoles, C. J., Creutz, C. E., and Shulman, N. R.,** Evidence for control of serotonin secretion from human platelets by hydroxyl ion transport and osmotic lysis, *Proc. Natl. Acad. Sci. U.S.A.,* 74, 5295, 1977.

60. **Brown, E. M., Pazoles, C. J., Creutz, C. E., Aurbach, G. D., and Pollard, H. B.,** Role of anions in parathyroid hormone release from dispersed bovine parathyroid cells, *Proc. Natl. Acad. Sci. U.S.A.,* 75, 876, 1978.

61. **Brown, E. M., Hurwitz, S., and Aurbach, G. D.,** Preparation of viable isolated bovine parathyroid cells, *Endocrinology,* 99, 1582, 1976.

62. **Pollard, H. B., Pazoles, C. J., Creutz, C. E., Scott, J. H., Zinder, O., and Hotchkiss, A.,** An osmotic mechanism for exocytosis from dissociated chromaffin cells, *J. Biol. Chem.,* 259, 1114, 1984.

63. **Hampton, R. Y. and Holz, R. W.,** Effects of changes in osmolality on the stability and function of cultured chromaffin cells and the possible role of osmotic forces in exocytosis, *J. Cell Biol.,* 96, 1082, 1983.

64. **Holz, R. W., Senter, R. A., and Sharp, R. R.,** Evidence that the H^+ electrochemical gradient across membranes of chromaffin granules is not involved in exocytosis, *J. Biol. Chem.,* 258, 7506, 1983.

65. **Schuldiner, S., Fishkes, H., and Kanner, B. I.,** Role of a transmembrane pH gradient in epinephrine transport by chromaffin granule membrane vesicles, *Proc. Natl. Acad. Sci. U.S.A.,* 75, 3713, 1978.

66. **Korchak, H. M., Eisenstat, B. A., Hoffstein, S. T., Dunham, P. B., and Weissmann, G.,** Anion channel blockers inhibit lysosomal enzyme secretion from human neutrophils without affecting generation of superoxide anion, *Proc. Natl. Acad. Sci. U.S.A.,* 77, 2721, 1980.

67. **Hook, W. A. and Siraganian, R. P.,** Influence of anions, cations and osmolarity on IgE-mediated histamine release from human basophils, *Immunology,* 43, 723, 1981.

68. **Orci, L. and Malaisse, W.,** Single and chain release of insulin secretory granules is related to anionic transport at exocytotic sites, *Diabetes,* 29, 943, 1980.

69. **Somers, G., Sener, A., Devis, G., and Malaisse, W. J.,** The stimulus-secretion coupling of glucose-induced insulin release. XLV. The anionic-osmotic hypothesis for exocytosis, *Pflugers Arch. Eur. J. Physiol.,* 388, 249, 1980.

70. **Pace, C. S. and Smith, J. S.,** The role of chemiosmotic lysis in the exocytotic release of insulin, *Endocrinology,* 113, 964, 1983.

71. **Tamagawa, T. and Henquin, J.-C.,** Chloride modulation of insulin release, $^{86}Rb^+$ efflux, and $^{45}Ca^{2+}$ fluxes in rat islets stimulated by various secretagogues, *Diabetes,* 32, 416, 1983.

72. **Baker, P. F. and Knight, D. E.,** Calcium-dependent exocytosis in bovine adrenal medullary cells with leaky plasma membranes, *Nature (London),* 276, 620, 1978.

73. **Baker, P. F. and Knight, D. E.,** Gaining access to the site of exocytosis in bovine adrenal medullary cells, *J. Physiol. (Paris),* 76, 497, 1980.
74. **Baker, P. F. and Knight, D. E.,** Calcium control of exocytosis and endocytosis in bovine adrenal medullary cells, *Phil. Trans. R. Soc. Lond. Ser. B.,* 296, 83, 1981.
75. **Knight, D. E. and Baker, P. F.,** Calcium-dependence of catecholamine release from bovine adrenal medullary cells after exposure to intense electric fields, *J. Membrane Biol.,* 68, 107, 1982.
76. **Knight, D. E. and Baker, P. F.,** The chromaffin granule proton pump and calcium-dependent exocytosis in bovine adrenal medullary cells, *J. Membrane Biol.,* 83, 147, 1985.
77. **Dunn, L. A. and Holz, R. W.,** Catecholamine secretion from digitonin-treated adrenal medullary chromaffin cells, *J. Biol. Chem.* 258, 4989, 1983.
78. **Wilson, S. P. and Kirshner, N.,** Calcium-evoked secretion from digitonin-permeabilized adrenal medullary chromaffin cells, *J. Biol. Chem.,* 258, 4994, 1983.
79. **Pace, C. S., Tarvin, J. T., Neighbors, A. S., Pirkle, J. A., and Greider, M. H.,** Use of a high voltage technique to determine the molecular requirements for exocytosis in islet cells, *Diabetes,* 29, 911, 1980.

Chapter 13

MEMBRANE INTERACTIONS AND FUSION

Jan Wilschut

TABLE OF CONTENTS

I. INTRODUCTION*

The process of exocytosis is a general mechanism by which many enzymes and other proteins or peptides, hormones, and transmitter molecules are secreted from cells. It involves attachment of a secretory vesicle to the inner aspect of the cellular plasmalemma and subsequent fusion of the two closely apposed membranes, resulting in release of the contents of the secretory granule into the extracellular medium. There is extensive ultrastructural support for the concept of exocytotic secretion. Palade has described the formation of an initial "pentalaminar complex" composed of the membrane of the secretory vesicle and the plasma membrane in close apposition.[1] The pentalaminar complex subsequently thins to a single bilayer membrane and finally breaks. Biochemical evidence for secretion by exocytosis includes the observation that the components of the aqueous compartment of the secretory granules, but not their membranes, are released concomitantly, irrespective of molecular weight.[2]

Very little is known about molecular mechanisms involved in exocytotic membrane fusion, except for the fact that a rise of the intracellular free Ca^{2+} concentration commonly precedes secretion.[2-4] However, the site of Ca^{2+} action is not known and there is no compelling evidence to indicate that Ca^{2+} is involved in the actual fusion reaction. In fact, in several systems it has been demonstrated that secretion can be induced artificially without an increase of the intracellular free Ca^{2+} concentration,[5,6] suggesting that under physiological conditions Ca^{2+} does not act at the level of membrane fusion, but rather at some earlier stage of the process.

Fusion of biological membranes in general, and exocytotic membrane fusion in particular, occurs in a highly specific fashion. The secretory granule fuses only with the cellular plasma membrane or, in the extended process of "compound exocytosis", with the membrane of a previously fused granule. In exocrine cells fusion is even restricted to the apical domain of the plasma membrane. This high degree of selectivity implies that specific recognition mechanisms must play a crucial role in the interaction between secretory vesicle and plasma membrane. However, virtually nothing is known about molecular factors involved in this initial specific adhesion. Also with respect to the subsequent fusion, molecular mechanisms remain as yet largely obscure.

Much of our current knowledge regarding membrane-membrane interactions and fusion has been derived from investigation of fusion in artificial phospholipid vesicle (liposome) systems.[7-10] The use of liposomal model systems has several obvious advantages. The structure of liposomes is relatively simple. They can be made either multilamellar or unilamellar. The multilamellar liposomes consist of several concentric lipid bilayer membranes, while the unilamellar vesicles are composed of a single spherical bilayer separating an aqueous compartment from the surrounding medium. The composition of the membrane of the liposome and of the aqueous liposomal contents can be easily manipulated. A further advantage of the use of phospholipid vesicles is that their mutual interaction can be studied under well-defined and controllable conditions.

The limitations of the liposomal model system ought to be recognized as well. Interaction between protein-free lipid vesicles, by necessity, is nonspecific. Therefore, one cannot expect to mimic the initial specific adhesion step in the interaction between biological membranes. Also with respect to the actual fusion reaction the liposomal model system lacks specificity. Thus, the model systems may reveal, in rather general terms, the molecular requirements for lipid bilayer fusion, but conclusions derived from investigation of such systems cannot

---

*   Abbreviations: DPA: dipicolinic acid, *N*-NBD-PE: *N*-(7-nitro-2,1,3-benzoxadiazol-4-yl)phosphatidylethanolamine, *N*-Rh-PE: *N*-(lissamine Rhodamine B sulfonyl)phosphatidylethanolamine, CL: cardiolipin, PC: phosphatidylcholine, PE: phosphatidylethanolamine, PS: phosphatidylserine, RET: resonance energy transfer, SUV: small unilamellar vesicles, LUV: large unilamellar vesicles, and DLVO: Derjaguin-Landau-Verwey-Overbeek.

be directly extrapolated to biological membrane fusion. For example, the fusion-inducing capacity of Ca^{2+} in model systems consisting of negatively charged phospholipid vesicles is likely to be different from the probably much more specific function of Ca^{2+} in exocytosis.

A related limitation of the study of phospholipid vesicle fusion is the lack of control of the process in such model systems. Clearly, unless the stimulus to fusion (e.g., Ca^{2+}) is removed, the process will simply continue. This implies that one cannot normally expect to see postfusion stability of the fusion products. Therefore, it is imperative to employ methodologies that allow observation of the initial stages of the fusion process in a kinetic and quantitative manner. Such methodologies and their application will be discussed briefly in Section II.

With regard to the central theme of this volume, very little can be said with certainty about the energetics of biological membrane fusion. There are extensive data, however, on *interbilayer* forces, derived from studies on the interaction between lipid membranes in stacked multibilayer systems and lipid bilayers adsorbed onto a hydrophilic surface. These forces will be discussed in Section III and related to the interaction between phospholipid vesicles. It appears that phospholipid vesicles may well adhere or aggregate without the occurrence of direct molecular contact between the apposed bilayers and that the major barrier for such contact, and thus for fusion to occur, is a strong short-range hydration repulsion. It is likely that also with biological membranes, after the initial specific adhesion is established, hydration repulsion constitutes the major barrier for subsequent fusion. The first requirement for fusion, therefore, is the disruption of this "wall of water".

An additional, although related, requirement for fusion of phospholipid vesicles, that no doubt also pertains to fusion of biological membranes, is the creation of instabilities or defects in lipid packing, serving as the focal points for hydrophobic interaction and fusion. The occurrence of defects in the bilayer structure is largely governed by *intrabilayer* forces, which will be discussed in Section IV and related to the fusogenicity of phospholipid vesicles. An interesting aspect of the possible role of bilayer defects in fusion is that they may in fact be induced by the interaction between the apposed membranes, which would restrict their creation as focal points for fusion to the site of intermembrane contact. Finally, in section V, I will discuss briefly the possible involvement of membrane proteins in fusion and attempt to indicate how proteins might act such that the general requirements for fusion are met under the conditions of membrane fusion during exocytosis.

II. DETERMINATION OF THE KINETICS OF FUSION IN PHOSPHOLIPID VESICLE SYSTEMS

A. Methodologies

Various methodologies have been utilized in the study of fusion in phospholipid vesicle systems.[7] Among these are nuclear magnetic resonance (NMR) and electron spin resonance (ESR) spectroscopy, fluorescence techniques, differential scanning calorimetry, electron microscopy, gel filtration, and turbidity measurements. It is important to note that the criteria underlying these techniques, (such as vesicle aggregation, increase in vesicle size, mixing of bilayer lipids, or leakage of vesicle contents) do not always, in themselves, meet a rigorous definition of membrane fusion. For example, vesicle aggregation can be completely reversible and does not, in itself, provide evidence for fusion. Likewise, mixing of membrane constituents or increase in vesicle size may be the result of fusion, but can also be due to exchange or transfer of individual molecules. The most rigorous criterion for phospholipid vesicle fusion is the coalescence of the internal compartment of the vesicles and the concomitant stoichiometric mixing of membrane lipids. The fluorescence assays described below are based on this stringent criterion. In addition, they allow to assess the fusion reaction in a kinetic and quantitative fashion. As discussed above, because of the lack of control of the

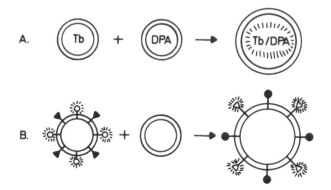

FIGURE 1. Schematic representation of the Tb/DPA assay (A) for mixing of aqueous contents and the RET assay (B) for mixing of bilayer lipids during fusion of phospholipid vesicles. In B, the triangles represent the donor probe molecules (*N*-NBD-PE) and the circles the acceptor probe molecules (*N*-Rh-PE).

fusion process in liposomal model systems, investigation of the very initial stages, preferably just the first dimerization, is imperative.

1. Mixing of Aqueous Vesicle Contents

The most commonly used assay for mixing of aqueous vesicle contents is the Tb/DPA assay (Figure 1A). Details of the assay have been described.[11-14] Briefly, it involves encapsulation of Tb^{3+} ions in one population of vesicles and the anion of DPA in another. To prevent the Tb^{3+} from binding to negatively charged phospholipids it is entrapped in the presence of a weak chelator, usually citrate. Fusion of Tb with DPA-containing vesicles results in the formation of the fluorescent $Tb(DPA)_3^{3-}$ complex. The formation of the complex is very fast, which permits an accurate determination of the kinetics of the fusion process. EDTA, included in the external medium, effectively prevents the formation of the fluorescent complex outside the vesicles, implying that any fluorescence observed is due to Tb/DPA complex formation sequestered from the external medium. With a modification of the Tb/DPA assay the kinetics of release of vesicle contents can be determined.[15]

2. Mixing of Bilayer Lipids

Bilayer lipid mixing can be monitored conveniently with an assay based on fluorescence RET between two derivatives of phosphatidylethanolamine, labeled in their polar groups: *N*-NBD-PE, the donor, and *N*-Rh-PE, the acceptor.[16,17] These fluorophores have been demonstrated to be nonexchangeable between phospholipid vesicles, even when the vesicles are aggregated.[16-19] The probes are incorporated together in one vesicle bilayer. Fusion of such labeled liposomes with unlabeled target membranes results in a decrease of the surface density of the probes and, hence, in a reduction of the energy transfer efficiency. The concomitant increase of the fluorescence of the donor (*N*-NBD-PE) is monitored continuously as a measure of fusion. Starting from probe concentrations in the vesicle bilayer of 0.6 to 0.7 mol% each, the relative fluorescence increase of *N*-NBD-PE is proportional to the dilution of the fluorophores, which permits a precise quantitation of the fusion process.[16,20] In addition to monitoring liposome-liposome fusion, this assay allows one to investigate fusion of liposomes with biological target membranes as well.

B. Ca^{2+}-Induced Fusion of Negatively Charged Phospholipid Vesicles

Vesicles composed of the negatively charged PS are stable in physiological salt solutions at neutral pH. Addition of millimolar concentrations of Ca^{2+} to PS vesicles causes vesicle

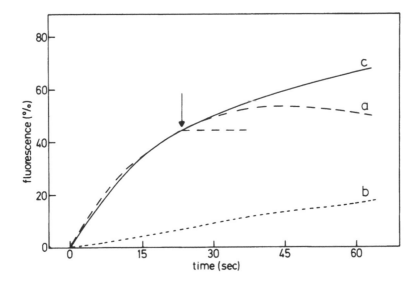

FIGURE 2.   Relative kinetics of mixing and release of vesicle contents and of mixing of bilayer lipids during Ca^{2+}-induced fusion of LUV composed of a 1:1 mixture of bovine heart CL and dioleoyl-PC. Final Ca^{2+} concentration was 11 mM. Curve a, mixing of aqueous vesicle contents (Tb/DPA assay); curve b, release of vesicle contents (modification of the Tb/DPA assay); curve c, mixing of bilayer lipids (RET assay). The fluorescence is calibrated such that the curves can be compared directly. The secondary decrease of curve a is due to release of vesicle contents. The arrow indicates (in a separate Tb/DPA experiment) the interruption of the fusion process by addition of excess EDTA. (From Wilschut, J. et al., *Biochemistry*, 24, 4630, 1985. With permission.)

aggregation, release of aqueous vesicle contents, and eventual formation of large cylindrical structures, so-called "cochleates", which transform into large unilamellar vesicles upon addition of excess EDTA.[21,22]

Obviously, the release of aqueous vesicle contents is disturbing in that it indicates the loss of vesicular integrity as a result of Ca^{2+}-induced vesicle-vesicle interaction. It is not surprising, therefore, that this observation has been the basis for criticism, questioning the validity of the Ca^{2+}/PS system as an appropriate model for fusion.[23-25] Rather than modeling membrane fusion, the process observed would merely represent a reassembly of bilayer fragments following the contact-induced rupture of the vesicles.

In order to distinguish between "fusion" and "rupture", the relative kinetics of aqueous contents mixing and release of vesicle contents during Ca^{2+}-induced fusion of negatively charged phospholipid vesicles have been analyzed extensively, employing the kinetic assays described above. Figure 2 shows the case of large unilamellar vesicles made of an equimolar mixture of CL and PC. Clearly, the mixing of aqueous vesicle contents (curve a) precedes their release into the external medium (curve b), indicating that, after aggregation, the vesicles communicate their internal contents in a relatively nonleaky fashion.[26,27] Similar results have been obtained with PS vesicles[11-15,29] and we have reported on several cases of fusion where leakage does not occur at all, not even eventually.[14]

Figure 2 also shows results on the mixing of bilayer lipids.[27] The initial kinetics of this process (curve c) are identical to those of the mixing of aqueous vesicle contents (curve a). Thus, it appears that fusion in the CL/PC vesicle system meets a most rigorous criterion for fusion, the nonleaky coalescence of internal aqueous compartments and a concomitant stoichiometric mixing of bilayer lipids.

The initial fusion events being largely nonleaky, why does extensive release of contents occur eventually? The answer to this question, in fact, is simple. Addition of Ca^{2+} to CL/PC or PS vesicles induces a transformation from an initial condition in which the lipid is present as stable bilayer vesicles to another equilibrium condition in which the lipid is in the form of giant nonvesicular aggregates. If this transformation is allowed to proceed in an entirely uncontrolled fashion, obviously, somewhere along the way the internal contents of the vesicles are lost. One may speculate that it is the result of collapse of fusion products.[12-14,21] Irrespective of the precise mechanism of release, it is important to emphasize that it is a secondary process due to the lack of control in liposomal model systems. If the stimulus to the fusion reaction (i.e. Ca^{2+}) is taken away, the process is interrupted and the aqueous vesicle contents are largely retained in the fused vesicles. This can be accomplished by addition of EDTA (Figure 2, arrow), which results in dissociation of aggregated vesicles and a fixation of the Tb fluorescence intensity, indicating an effective sequestration of the Tb/DPA complex from the external medium. Interruption of the fusion process parallels biological membrane fusion in the sense that these events are often induced by stimuli that are only transiently present.

III. INTERBILAYER FORCES: VESICLE AGGREGATION AND ESTABLISHMENT OF MOLECULAR CONTACT

Obviously, a necessary condition for fusion between membranes to occur is the establishment of mutual contact. As discussed above, the interaction between biological membranes that are about to fuse is highly specific and must rely on specialized recognition mechanisms. Molecular components involved in this initial, specific interaction have not been identified yet. One may speculate on a crucial role for carbohydrate-containing proteins and/or lipids in this respect.

While little is known about specific adhesion mechanisms between biological membranes, knowledge on the interaction between lipid bilayers and the forces involved is quite extensive. This knowledge is based not only on studies regarding the aggregation behavior of phospholipid vesicles, but also on direct measurement of forces between lipid bilayer membranes and observation of their structure as a function of bilayer separation. It should be recognized that the nature of the intermembrane interaction in these systems is different from the specific adhesion between biological membranes. Yet, the information derived from studies on interacting phospholipid bilayers is highly relevant for biological membrane interactions, since at some stage of the physiological fusion process the core structures of the two membranes, the lipid bilayers, must come into molecular contact and merge. It is likely that during the establishment of this molecular contact, as opposed to the initial adhesion, the interaction between biological membranes is governed by the same forces that are operating between model bilayer membranes.

As two lipid bilayer membranes approach one another they experience different interaction forces.[30-33] These forces fall into distinct categories. They can either be long range, involving attraction or repulsion across an intervening aqueous space, or short range, involving direct molecular interactions. Clearly, it is the short-range interaction that is particularly relevant for biological membrane fusion. The prominent force operating in this regime is the so-called hydration repulsion, to be discussed below. However, first we will briefly discuss two other types of surface interaction, the electrostatic interaction and Van der Waals attraction.

A. Electrostatic Repulsion and Van der Waals Attraction

Electrostatic and Van der Waals interactions both represent long-range forces, i.e., they not only act at short distances, but also at separations between the interacting surfaces much greater than, say, 1 nm. The two forces are of contrasting character.[30-33] Electrostatic (or

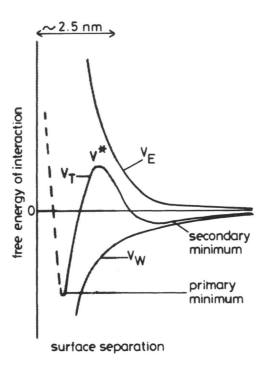

FIGURE 3. Free energy of interaction between two vesicles of like charge as a function of their surface separation, as described by the DLVO theory ignoring hydration repulsion. The total potential energy, V_T, is the sum of the electrostatic repulsion, V_E, and the Van der Waals attraction, V_W. The range of the repulsive hydration forces is approximately 2.5 nm, as indicated in the figure.

coulombic) interactions are due to the presence of fixed charges on the particle surface, giving rise to an electrostatic potential. Consequently, the interaction between particles of like charge is repulsive (Figure 3). It decays exponentially with increasing surface separation. Ions in the medium surrounding a charged particle distribute themselves in the region adjacent to the interface in response to the presence of charges on the particle surface. This screening is described by the classical Gouy-Chapman formulation of the diffuse double layer. The screening effect modifies the shape of the electrostatic surface potential. Thus, the rate of the exponential decay of coulombic repulsion between surfaces of like charge, as given by the Debye constant κ, increases with increasing ionic strength of the medium. At physiological ionic strength the Debye length, $1/\kappa$, is about 0.9 nm.

In contrast to electrostatic interaction, the Van der Waals (or electrodynamic) force between two like particles is attractive in nature (Figure 3). It arises from charge fluctuations in the interacting particles and the surrounding medium, which are reflected in their respective polarizabilities. The Van der Waals force between large bodies in condensed media falls off far more slowly than with the classical inverse sixth power of the distance of separation, which does apply to interaction between isolated atoms in vapor. In fact, Van der Waals attraction may decay even more slowly than electrostatic repulsion. It is for this reason that Van der Waals forces, acting at long range, can stabilize aggregates of particles under conditions where no direct surface contact occurs (Figure 3, see also below).

B. Hydration Forces

When phospholipid bilayers are forced to close approach strong repulsive hydration forces arise, which are due to the work involved in removal of the water that is tightly bound to the phospholipid headgroups.[33] The importance of hydration forces for intermembrane interaction and fusion can hardly be overestimated. They have been discovered by direct investigation of the interaction energies between phospholipid bilayers as a function of their mutual separation. Rand et al.[33-35] have determined forces between phospholipid bilayers in stacked multibilayer systems, while observing interbilayer separation by X-ray diffraction.[33-35] In this system bilayers can be forced to close approach in various ways, including osmotic swelling or application of mechanical pressure. Marra and Israelachvili[36] and Horn[37] have measured forces between phospholipid bilayers adsorbed onto mica surfaces. Results obtained with these different methodologies are similar. Hydration forces start to arise at surface separations of about 2 to 3.5 nm, depending on the nature of the lipid (Figure 3). With decreasing distance of separation the force rises steeply. In this regime of surface separations hydration repulsion is the dominating interactive force, largely overwhelming both electrostatic repulsion and Van der Waals attraction.

Hydration repulsion occurs between phospholipid bilayers of any composition, but its magnitude may vary with the nature of the particular lipid involved. For example, for neutral PC bilayers the equilibrium separation, where hydration repulsion is balanced by Van der Waals attraction, is about 2.1 to 2.5 nm.[33-37] By contrast, for PE, which is also zwitterionic, the equilibrium separation distance is about 1 nm smaller and the adhesive energy almost an order of magnitude larger.[33,35,36,38] This difference is due to the different conformation and the lower degree of hydration of PE, as compared to that of PC.[39,40]

Finally, it should be noted that hydration repulsion is not due solely to effects of surface-structured water. The interfacial aqueous region is not pure water but also contains the thermally mobile phospholipid headgroups. These contribute a complex steric component to the short-range repulsive force. The steric contribution to the total repulsion is affected by fluctuations in bilayer thickness and by bilayer undulations. This is probably why the measured hydration repulsion between bilayers adsorbed onto mica surfaces[36,37] is of somewhat shorter range than that in multibilayer systems,[33-35] since the former will have restricted possibilities for undulation. Furthermore, it explains the observed extended range of hydration repulsion between "fluid" bilayers, relative to "solid" bilayers, as an effect of increased thermal bilayer fluctuations and undulations.[36]

C. Aggregation of Phospholipid Vesicles

A local minimum in the free energy of interaction between phospholipid vesicles is a sufficient condition for aggregation of the vesicles to occur. The depth of the energy well determines the average period of time the vesicles remain in the aggregated state. A well with a depth in the order of kT (where k is Boltzmann's constant and T the absolute temperature) or less is too shallow to support measurable aggregation.

There is ample evidence indicating that a theory on the interaction between lyophobic colloidal particles in most cases provides an adequate description of the aggregation of phospholipid vesicles. This theory, the DLVO theory, takes into account electrostatic repulsion and Van der Waals attraction.[41] It predicts for the interaction between charged particles that there are two types of energy minimum, depending on the distance of particle separation. The primary minimum occurs at surface separations of 1 nm or less and the secondary minimum at separations of 3 to 10 nm (Figure 3). Aggregation in the secondary minimum has been demonstrated for negatively charged PS vesicles under certain ionic conditions.[42] However, rather than this type of aggregation, obviously, aggregation in the primary minimum, at much closer surface separations, is of relevance to bilayer fusion. The rate of aggregation in the primary minimum is determined by the height of the potential energy barrier, V*, which, in turn, depends on the interplay between electrostatic repulsion and Van der Waals attraction (Figure 3[7]). In a medium containing physiological salt concentrations vesicles composed of negatively charged phospholipids, such as PS, will not aggregate due to electrostatic repulsion: the energy barrier, V*, is too high. Monovalent and divalent cations induce aggregation of the vesicles primarily by direct binding to the phospholipid headgroups, resulting in charge neutralization. Cations also screen the charge on the vesicle surface and, thereby, cause an additional lowering of the energy barrier. For example, SUV made of PS aggregate in the presence of Na^+ concentrations of 550 mM or higher or in 100 mM Na^+ in combination with 1 mM Ca^{2+} or 4 mM Mg^{2+}.[42-45] These observations are in perfect agreement with predictions based on the DLVO theory, provided that the physical binding of the cations to PS is taken into account.[7,32,42] Thus, the threshold cation concentrations at which the vesicles start to aggregate are determined by the binding affinities of cations for the anionic lipid. These affinities are much higher for divalent cations than for monovalent ones, explaining the difference between the threshold concentrations of Na^+ and Ca^{2+} required for aggregation of PS vesicles.

Aggregation of PS vesicles in high concentrations of Na^+ is less extensive than aggregation induced by Ca^{2+}. This is a reflection of the dynamic nature of the aggregation process.[43,44] In the presence of Ca^{2+} it appears largely irreversible, since the aggregated vesicles fuse almost immediately, i.e., they do not get a chance to dissociate. However, in NaCl, where fusion does not occur, the process is highly reversible involving aggregation and dissociation.[43] This aggregation occurs in the primary energy minimum (Figure 3), since with SUV the secondary minimum under these conditions is too shallow to support measurable aggregation.[32-42] Therefore, the vesicles can escape from the primary energy well. This results in an increased degree of dispersion at higher temperatures.[43]

As noted above, the DLVO theory does not take into account short-range hydration repulsion. As the theory provides an adequate description of the aggregation behavior of anionic phospholipid vesicles, this could be taken to suggest that hydration forces have no effect on vesicle aggregation. Straightforward extrapolation of the conclusions on hydration repulsion, as discussed above, would imply that the vesicles could approach each other no closer than, say, 2 nm. Indeed, PC SUV in the "fluid" state do not aggregate appreciably. Obviously, there is no potential barrier between these zwitterionic phospholipid vesicles and there is only one energy minimum. It is likely that this minimum is too shallow to support aggregation due to the effect of hydration repulsion.[33,36,38] As discussed above, the energy minimum between PE bilayers is deeper, because of the relatively low degree of hydration of PE. It is, at least in part, for that reason that PE vesicles do have a tendency to aggregate.[46] Therefore, it would appear that, indeed, hydration repulsion affects the aggregation of zwitterionic phospholipid vesicles. For vesicles consisting of anionic phospholipids, such as PS, a 2-nm range of the hydration repulsion would imply that aggregation in the primary minimum would be impossible (Figure 3). However, PS SUV do aggregate in the primary minimum in the presence of, for example, Na^+ at concentrations of 550 mM or higher.[43] This means that either the range of hydration repulsion between PS bilayers does not extend beyond about 1 nm or that this range is shorter between small vesicles than between extended flat bilayers or both. There is evidence indicating that, in terms of hydration, PS more closely resembles PE than PC.[33] In any case, aggregation of anionic phospholipid vesicles is consistent with predictions of the DLVO theory, ignoring hydration repulsion. It is likely that, in the overall process of fusion, hydration repulsion has its prominent effect in the stage *after* vesicle aggregation, i.e., during the establishment of molecular contact.

A final aspect affecting vesicle aggregation is the degree of "deformability" or "flaccidity" of the vesicles.[38,47] When two vesicles adhere to one another they will tend to enlarge their area of contact. By increasing the area of contact, flattening deformation enhances the net attraction between the vesicles. However, deformation also induces a counteracting tension on the membrane. In rigid vesicles, this tension will build up very rapidly. This is the case for "fluid" SUV, for example, and, therefore, these small vesicles will form a limited area of contact and, thus, a less stable association. Larger, flaccid, vesicles can form more extended areas of contact with little work, such that their association is strengthened. Contact-induced vesicle deformation and the concomitant development of membrane tension may be one of the factors leading to bilayer destabilization, perturbation of lipid packing, and fusion.

D. Molecular Contact: Hydration Forces Pose a Major Barrier

Fusion of two lipid bilayers requires the establishment of direct molecular contact. Several lines of evidence indicate that it is at this stage, more than during the formation of vesicle aggregates, that hydration repulsion poses a major barrier.

First, the capacity of poly(ethyleneglycol) (PEG) to induce fusion of phospholipid vesicles points to a key role of bilayer dehydration in the induction of direct interbilayer contact.[48]

Second, freezing and thawing cause dehydration of phospholipid bilayers and have been demonstrated to induce fusion as well.[49]

Third, the lipid specificity of phospholipid vesicle fusion also indicates a major role of interfacial hydration in modulating the establishment of direct intervesicle contact. In several mixed phospholipid vesicle systems, consisting of anionic and zwitterionic phospholipids, fusion capacity is sustained or sometimes even enhanced, when PE is the zwitterionic component.[50-52] On the other hand, PC is almost always strongly inhibitory. For example, LUV consisting of a 1:1 mixture of PS and PC just aggregate in the presence of Ca^{2+}, but have lost their capacity to fuse.[50] Also in SUV systems PC causes a strong inhibition (Figure 4). This inhibition is likely to be due to the strong hydration of the polar group of PC, preventing the establishment of direct interbilayer contact.

Fourth, the cation specificity of PS vesicle fusion correlates with the degree to which the ions are dehydrated and with their ability to form dehydrated complexes with PS.[53-54] The most extreme case is given by Mg^{2+} which does not induce any fusion of PS LUV, even though the vesicles aggregate massively.[13] Figure 5B shows PS LUV that have been aggregated in the presence of Mg^{2+} and subsequently dispersed by treatment with excess EDTA. The average size of the vesicles is the same as that of control vesicles (Figure 5A), while Ca^{2+}-treated vesicles (Figure 5C) show a large increase in size. It has been demonstrated that Ca^{2+} has the capacity to completely dehydrate the space between two PS bilayers, presumably by forming a *trans* complex involving PS molecules on the apposed surfaces.[45] On the other hand, with Mg^{2+} a *cis* complex is formed and a layer of water remains between the bilayers, thus preventing direct interbilayer contact and fusion.[45] The formation of a *trans* Ca^{2+}/PS complex is indicated by the observation that the binding of Ca^{2+} to PS is greatly enhanced when the vesicles aggregate and fuse.[55]

IV. INTRABILAYER FORCES: MEMBRANE DESTABILIZATION AND FUSION

A. Local Defects in Lipid Packing

From the discussion in the preceding section it is evident that the formation of a dehydrated area of interbilayer contact is a necessary condition for fusion to occur. However, it is not sufficient. A local perturbation of lipid packing, serving as a focal point for fusion, is required as well. In experiments involving phospholipid bilayers adsorbed onto mica surfaces, Marra and Israelachvili[36] and Horn[37] have observed occasional fusion events under conditions of extremely large interbilayer pressures. Fusion was initiated at one point via a local thinning of the closely apposed and largely dehydrated bilayers. Interestingly, it was noted that the shape of the force law between the interacting bilayers down to the point of fusion was not obviously different from other force laws when there was no fusion.[36] This indicates that fusion is not governed solely by interbilayer forces, but that *intra*bilayer forces causing a

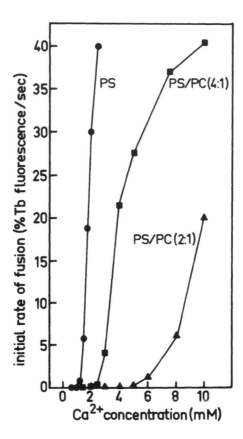

FIGURE 4. Inhibition of Ca^{2+}-induced PS SUV fusion by PC. Fusion rates were determined with the Tb/DPA assay for SUV of the compositions indicated. (Wilschut, J. and Düzgünes, N., unpublished observations.)

local membrane destabilization must also be considered. Horn noted that fusion of bilayers adsorbed onto mica surfaces is facilitated by the presence of impurities in the lipid, causing imperfections in the bilayer structure.[37] It seems reasonable to assume that the role of membrane defects in promoting fusion lies in the exposure of the hydrophobic interior of the lipid bilayer. This would permit hydrophobic interactons, the magnitude of which may be quite large.[56] The concept of bilayer fusion proceeding through point defects has been proposed by Hui et al.[49] It is also consistent with the theory on membrane fusion of Ohki, emphasizing the importance of increased surface hydrophobicity.[57]

The role of lipid packing defects in bilayer fusion is indicated by the intrinsic tendency of SUV to fuse more readily than larger vesicles.[7,12-14,29,54,58,59] The high degree of curvature of the lipid bilayer in SUV creates constraints in the packing of the lipid molecules. These packing imperfections make the bilayer of SUV particularly susceptible to destabilization. In this respect it is interesting to note that Mg^{2+}, while unable to induce fusion of PS LUV (Figure 5B), does induce fusion of PS SUV (Figure 5E). Fusion in this case proceeds to a limited extent: as soon as the strain imposed on the vesicle bilayer is relieved, as a result of the increase in vesicle size, the fusion process stops spontaneously.[13]

Fusion of PS SUV has been demonstrated to be greatly facilitated by PEG, but it depends critically on the presence of divalent cations.[48] This indicates that, also in the case of PEG-promoted fusion, an additional bilayer destabilization is required. PEG-induced cell fusion requires osmotic swelling of the cells in addition to the close membrane apposition resulting from surface dehydration.[60]

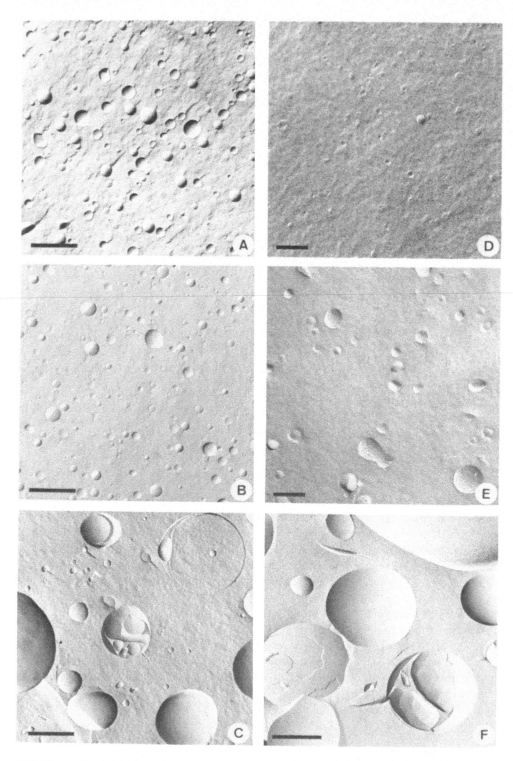

FIGURE 5. Freeze-fracture electron micrographs of PS vesicles before and after interaction with Ca²⁺ or Mg²⁺. Vesicles were incubated with CaCl₂ or MgCl₂ for 10 min at 25°C. The vesicle aggregates were spun down and, to disaggregate the vesicles, the pellets were treated with excess EDTA. (A) LUV control; (B) LUV 10 m*M* Mg²⁺; (C) LUV 5 m*M* Ca²⁺; (D) SUV control; (E) SUV 8 m*M* Mg²⁺; (F) SUV 3 m*M* Ca²⁺. Bars represent 0.5 μm in A—C and F and 0.1 μm in D and E. (From Wilschut, J. et al., *Biochemistry*, 20, 3126, 1981. With permission.)

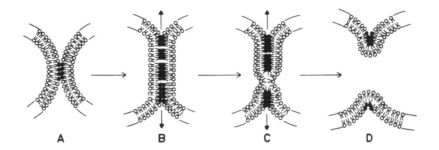

FIGURE 6. Hypothetical mechanism of divalent cation-induced fusion of anionic phospholipid vesicles. (A) vesicle aggregation and the formation of a dehydrated point of contact. (B) Expansion of the area of contact results in an increased surface to volume ratio of the vesicles and generates a tension on the vesicle bilayers; lipid packing defects develop at the area of contact. (C) A point defect in the apposed bilayers and the deformation-induced tension on the vesicles produce a local thinning of the diaphragm. (D) Breakage and bilayer merging; it is unlikely that the fused structure will swell to such an extent that a perfect sphere is formed.

The formation of defects in the bilayer structure is likely to be a local event, occurring only at the site of intermembrane contact. For example, in the Ca^{2+}/PS system the formation of a *trans* complex involving PS molecules in the apposing bilayers, as discussed in the previous section, is a local event that, obviously, requires interbilayer contact.[45,55] The formation of the *trans* complex presumably induces local perturbations of the lipid packing (Figure 6B), allowing direct hydrophobic interactions. Another example of a contact-induced destabilization of lipid bilayers is given by the interaction between vesicles consisting of unsaturated PE, as studied recently by Bentz et al.[61,62] They found that the susceptibility of the vesicles to destabilization is greatly enhanced under conditions where the PE species used prefers a hexagonal (H_{II}) arrangement rather than a bilayer structure. Thus, the polymorphic phase behavior of PE provides a powerful way of destabilizing membranes in a contact-dependent fashion.

B. Contact-Induced Vesicle Deformation and the Role of Osmotic Forces

As discussed in the previous section, when two vesicles adhere they have the tendency, depending on their interactive energy, to enlarge the effective area of contact. The result of increasing the area of contact is a flattening deformation of the vesicles, which produces a tension on the membrane.[38,47] If the deformation is extensive this will eventually result in rupture of the membrane,[63] since by flattening of the vesicles their surface to volume ratio increases, which, in turn, produces an osmotic stress on the vesicles. Rand and co-workers[25,64] have argued that this deformation-induced membrane rupture causes lysis as the predominant event in the interaction among PS vesicles in the presence of Ca^{2+}. However, on the basis of extensive data, demonstrating that after Ca^{2+}-induced PS vesicle aggregation the vesicles initially communicate their contents in a relatively nonleaky fashion,[11-14] concomitant with the occurrence of lipid mixing,[28] I tend to conclude that the vesicle bilayers preferentially break at the site of the diaphragm due to the presence of local defects in lipid packing.[27,45,49] Thus, the osmotic strain imposed on the vesicles as a result of the flattening deformation facilitates the final merging of the bilayers (Figure 6). Within this concept the degree of bilayer destabilization at the site of interaction determines to what extent the vesicles will have to be deformed before merging of their bilayers occurs. In other words, if there are severe local packing defects, one would expect that little, if any, osmotic strain would suffice to induce fusion. On the other hand, in case there are only minor lipid perturbations a significant vesicle deformation and osmotic stretching may be required. In this respect it is interesting to note that osmotic strain has been demonstrated to be an absolute prerequisite for fusion of phospholipid vesicles with planar bilayers.[65-67] Moreover, osmotic strain as a result of an applied osmotic gradient has been shown to facilitate fusion among phospholipid vesicles.[68,69]

V. CONCLUSIONS AND POSSIBLE IMPLICATIONS FOR EXOCYTOTIC MEMBRANE FUSION

Phospholipid vesicle systems have proven to be of great value in the study of the interactions between lipid bilayer membranes. On the basis of the investigation of various model membrane systems we now have a reasonably coherent picture of the general requirements of membrane fusion. These requirements involve a local dehydration of the membrane surface and perturbations of the lipid packing at the site of intermembrane contact, constituting the focal points for fusion. Once more I emphasize the *general* character of these requirements. The knowledge derived from a particular phospholipid vesicle system cannot be extrapolated directly to a seemingly comparable counterpart in biology. For example, as already noted in Section I, the effect of Ca^{2+} on anionic phospholipid vesicles does not necessarily pertain to the role of Ca^{2+} in exocytosis. Although the possibility of a direct Ca^{2+}-phospholipid interaction cannot be excluded altogether, it seems likely that soluble cytosolic and membrane-associated proteins are critically involved in exocytotic membrane fusion.

With respect to the role of proteins in membrane fusion, early freeze-fracture studies have revealed the formation of "bare lipid patches" in fusing membranes. This has led to the suggestion that proteins are not involved in the actual fusion process. Rather, in this concept they had to be cleared from the site of intermembrane contact, where lipid-lipid interactions would suffice to bring about fusion.[70] However, recent freeze-fracture studies, in which ultrafast freezing techniques were employed such that the use of cryoprotectants could be avoided, have demonstrated that membrane proteins need not be cleared at all from the adhesion site between fusing membranes, suggesting a direct involvement of proteins in the fusion process.[71-73]

The fusion-inducing capacities of several soluble peptides or proteins have been investigated in various model membrane systems. Examples include polymyxin B,[74] melittin,[75] tubulin,[19] clathrin,[76] bovine serum albumin[77] or albumin fragments,[78] and lysin, a cationic protein from spermatozoa.[79] A common feature seems to emerge from these studies indicating that the peptide or protein needs to be amphipathic in nature in order to have the ability to promote lipid bilayer fusion. The amphipathic character of the molecules implies that, at least under conditions where fusion occurs, one or more hydrophobic amino acid sequences are exposed, which, in principle, share the capacity to penetrate into the hydrophobic interior of a lipid bilayer. This is suggestive of a mechanism of fusion in line with the conclusions derived from the work on ion-induced fusion, since penetration of a hydrophobic peptide segment into a lipid bilayer can be expected to disrupt the interfacial hydration layer and to cause defects in lipid packing.

An example of a membrane-associated fusion protein that is characterized in considerable detail is given by the hemagglutinin of influenza virus. The current view on the way this protein induces membrane fusion is entirely consistent with the general conclusions derived from the liposomal model systems. Thus, the capacity of the protein to induce fusion is thought to be activated by a pH-dependent conformational change in the molecule, exposing a hydrophobic amino acid sequence, which would induce fusion through penetration into the target membrane.[80,81] One may speculate on a similar involvement of membrane proteins in exocytotic fusion events (Figure 7). Recent ultrastructural evidence indicates the presence of membrane proteins at the site of fusion during exocytosis in several different cell types.[71-73] The fusion reaction does not appear to involve the formation of an extended diaphragm, as suggested by Palade,[1] but rather occurs in a small area (probably less than

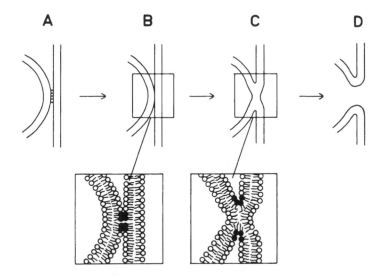

FIGURE 7. Hypothetical mechanism of exocytotic membrane fusion. (A) Specific adhesion of the secretory granule to the plasma membrane. (B) Local dehydration and establishment of direct molecular contact. (C) Perturbation of lipid packing and thinning of the "pentalaminar" structure. (D) Breakage and formation of the exocytotic canal. It is likely that in stages A through C of the process soluble cytosolic as well as membrane-associated proteins play a key role, while in C and D osmotic swelling of the secretory granule may be involved.

10 nm in diameter) consistent with the concept of a local perturbation in lipid packing serving as the nucleation point for fusion.[72,73] Osmotic swelling of the secretory vesicle[67,82] (see also Chapter 12 in this volume) may facilitate the destabilization of the apposed membranes at the site of contact.

ACKNOWLEDGMENTS

I wish to thank my collaborators in the work cited in this chapter, in particular Drs. Dick Hoekstra, Nejat Düzgünes, Shlomo Nir, and Demetrios Papahadjopoulos. Special thanks to Mrs. Janny Scholma for her contribution to this work, to Dr. Gerrit Scherphof for critically reading the manuscript and to Mrs. Rinske Kuperus for expert secretarial assistance.

I acknowledge the financial support from The Netherlands Organization for the Advancement of Pure Research (Z.W.O.), NATO (Research Grant 151.81), and EMBO.

REFERENCES

1. **Palade, G.,** Intracellular aspects of the process of protein synthesis, *Science,* 189, 347, 1975.
2. **Pollard, H. B., Pazoles, C. J., Creutz, C. E., and Zinder, O.,** The chromaffin granule and possible mechanisms of exocytosis, *Int. Rev. Cytol.,* 58, 159, 1979.
3. **Baker, P. F. and Knight, D. E.,** Calcium control of exocytosis in bovine adrenal medullary cells, *Trends Neurosci.,* 7, 120, 1984.
4. **Burgoyne, R. D.,** Mechanisms of secretions from adrenal chromaffin cells, *Biochim. Biophys. Acta,* 779, 201, 1984.
5. **Rink, T. J., Sanchez, A., and Hallam, T. J.,** Diacylglycerol and phorbol ester stimulate secretion without raising cytoplasmic free calcium in human platelets, *Nature (London),* 305, 317, 1983.

6. **Korchak, H. M., Vienne, K., Rutherford, L. E., Wilkenfeld, C., Finkelstein, M. C., and Weissmann, G.,** Stimulus-response coupling in the human neutrophil. II. Temporal analysis of changes in cytosolic calcium and calcium efflux, *J. Biol. Chem.,* 259, 4076, 1984.

7. **Nir, S., Bentz, J., Wilschut, J., and Düzgünes, N.,** Aggregation and fusion of phospholipid vesicles, *Prog. Surf. Sci.,* 13, 1, 1983.

8. **Wilschut, J. and Hoekstra, D.,** Membrane fusion: from liposomes to biological membranes, *Trends Biochem. Sci.,* 9, 479, 1984.

9. **Düzgünes, N.,** Membrane fusion, in *Subcellular Biochemistry,* Vol. 11, Roodyn, D. B., Ed., Plenum Press, New York, 1985, 195.

10. **Wilschut, J. and Hoekstra, D.,** Membrane fusion: lipid vesicles as a model system, *Chem. Phys. Lipids,* 40, 145, 1986.

11. **Wilschut, J. and Papahadjopoulos, D.,** Ca^{2+}-induced fusion of phospholipid vesicles monitored by mixing of aqueous contents, *Nature (London),* 281, 690, 1979.

12. **Wilschut, J., Düzgünes, N., Fraley, R., and Papahadjopoulos, D.,** Studies on the mechanisms of membrane fusion: kinetics of calcium ion-induced fusion of phosphatidylserine vesicles followed by a new assay for mixing of aqueous vesicle contents, *Biochemistry,* 19, 6011, 1980.

13. **Wilschut, J., Düzgünes, N., and Papahadjopoulos, D.,** Calcium/magnesium specificity in membrane fusion: kinetics of aggregation and fusion of phosphatidylserine vesicles and the role of bilayer curvature, *Biochemistry,* 20, 3126, 1981.

14. **Wilschut, J., Düzgünes, N., Hong, K., Hoekstra, D., and Papahadjopoulos, D.,** Retention of aqueous contents during divalent cation-induced fusion of phospholipd vesicles, *Biochim. Biophys. Acta,* 734, 309, 1983.

15. **Bentz, J., Düzgünes, N., and Nir, S.,** Kinetics of divalent cation-induced fusion of phosphatidylserine vesicles: correlation between fusogenic capacities and binding affinities, *Biochemistry,* 22, 3320, 1983.

16. **Struck, D. K., Hoekstra, D., and Pagano, R. E.,** Use of resonance energy transfer to monitor membrane fusion, *Biochemistry,* 20, 4093, 1981.

17. **Hoekstra, D.,** Role of lipid phase separations and membrane hydration in phospholipid vesicle fusion, *Biochemistry,* 21, 2833, 1982.

18. **Nichols, J. W. and Pagano, R. E.,** Resonance energy transfer assay of protein-mediated lipid transfer between vesicles, *J. Biol. Chem.,* 258, 5368, 1983.

19. **Kumar, N., Blumenthal, R., Henkart, M., Weinstein, J. N., and Klausner, R. D.,** Aggregation and calcium-induced fusion of phosphatidylcholine vesicle-tubulin complexes, *J. Biol. Chem.,* 257, 15137, 1982.

20. **Driessen, A. J. M., Hoekstra, D., Scherphof, G., Kalicharan, R. D., and Wilschut, J.,** Low pH-induced fusion of liposomes with membrane vesicles derived from *Bacillus subtilis, J. Biol. Chem.,* 260, 10880, 1985.

21. **Papahadjopoulos, D., Vail, W. J., Jacobson, K., and Poste, G.,** Cochleate lipid cylinders: formation by fusion of unilamellar lipid vesicles, *Biochim. Biophys. Acta,* 394, 483, 1975.

22. **Papahadjopoulos, D., Vail, W. J., Newton, C., Nir, S., Jacobson, K., Poste, G., and Lazo, R.,** Studies on membrane fusion. III. The role of calcium-induced phase changes, *Biochim. Biophys. Acta,* 465, 579, 1977.

23. **Ginsberg, L.,** Does calcium cause fusion or lysis of unilamellar lipid vesicles? *Nature (London),* 275, 758, 1978.

24. **Kendall, D. A. and McDonald, R. C.,** A fluorescence assay to monitor vesicle fusion and lysis, *J. Biol. Chem.,* 257, 13892, 1982.

25. **Rand, R. P. and Parsegian, V. A.,** Physical force considerations in model and biological membranes, *Can. J. Biochem. Cell Biol.,* 62, 752, 1984.

26. **Wilschut, J., Holsappel, M., and Jansen, R.,** Ca^{2+}-induced fusion of cardiolipin/phosphatidylcholine vesicles monitored by mixing of aqueous vesicle contents, *Biochim. Biophys. Acta,* 690, 297, 1982.

27. **Wilschut, J., Nir, S., Scholma, J., and Hoekstra, D.,** Kinetics of Ca^{2+}-induced aggregation and fusion of cardiolipin/phosphatidylcholine vesicles: correlation between vesicle aggregation, bilayer destabilization and fusion, *Biochemistry,* 24, 4630, 1985.

28. **Wilschut, J., Scholma, J., Bental, M., Hoekstra, D., and Nir, S.,** Ca^{2+}-induced fusion of phosphatidylserine vesicles: mass action kinetic analysis of membrane lipid mixing and aqueous contents mixing, *Biochim. Biophys. Acta,* 821, 45, 1985.

29. **Wilschut, J., Düzgünes, N., Hoekstra, D., and Papahadjopoulos, D.,** Modulation of membrane fusion by membrane fluidity: temperature dependence of divalent cation-induced fusion of phosphatidylserine vesicles, *Biochemistry,* 24, 8, 1985.

30. **Parsegian, V. A.,** Long-range physical forces in the biological milieu, *Annu. Rev. Biophys. Bioeng.,* 2, 221, 1973.

31. **Nir, S.,** Van der Waals interactions between surfaces of biological interest, *Prog. Surf. Sci.,* 8, 1, 1977.

32. **Nir, S. and Bentz, J.,** On the forces between phospholipid bilayers, *J. Colloid Interface Sci.*, 65, 399, 1978.
33. **Rand, R. P.,** Interacting phospholipid bilayers: measured forces and induced structural changes, *Annu. Rev. Biophys. Bioeng.*, 10, 277, 1981.
34. **LeNeveu, D. M., Rand, R. P., and Parsegian, V. A.,** Measurement of forces between lecithin bilayers, *Nature (London)*, 259, 601, 1976.
35. **Lis, L. J., McAlister, M., Fuller, N., Rand, R. P., and Parsegian, V. A.,** Interactions between neutral phospholipid bilayer membranes. *Biophys. J.*, 37, 657, 1982.
36. **Marra, J. and Israelachvili, J.,** Direct measurements of forces between phosphatidylcholine and phosphatidylethanolamine bilayers in aqueous electrolyte solution, *Biochemistry*, 24, 4608, 1985.
37. **Horn, R. G.,** Direct measurement of the force between two lipid bilayers and observation of their fusion, *Biochim. Biophys. Acta*, 778, 224, 1984.
38. **Parsegian, V. A. and Rand, R. P.,** Membrane interaction and deformation, *Ann. N.Y. Acad. Sci.*, 416, 1, 1983.
39. **Hauser, H., Pascher, I., Pearson, R. H., and Sundell, S.,** Preferred conformation and molecular packing of phosphatidylethanolamine and phosphatidylcholine, *Biochim. Biophys. Acta*, 650, 21, 1981.
40. **Jendrasiak, G. L. and Hasty, J. H.,** The hydration of phospholipids, *Biochim. Biophys. Acta*, 337, 79, 1974.
41. **Verwaey, E. J. and Overbeek, J. Th. G.,** *Theory of the Stability of Lyophobic Colloids*, Elsevier, Amsterdam, 1948.
42. **Nir, S., Bentz, J., and Düzgünes, N.,** Two modes of reversible vesicle aggregation: particle size and the DLVO theory, *J. Colloid Interface Sci.*, 84, 266, 1981.
43. **Day, E. P., Kwok, A. Y. W., Hark, S. K., Ho, J. T., Vail, W. J., Bentz, J., and Nir, S.,** Reversibility of sodium-induced aggregation of sonicated phosphatidylserine vesicles, *Proc. Natl. Acad. Sci. U.S.A.*, 77, 4026, 1980.
44. **Bentz, J. and Nir, S.,** Aggregation of colloidal particles modeled as a dynamical process, *Proc. Natl. Acad. Sci. U.S.A.*, 78, 1634, 1981.
45. **Portis, A., Newton, C., Pangborn, W., and Papahadjopoulos, D.,** Studies on the mechanism of membrane fusion: evidence for an intermembrane Ca^{2+}-phospholipid complex, synergism with Mg^{2+} and inhibition by spectrin, *Biochemistry*, 18, 780, 1979.
46. **Kolber, M. A. and Haynes, D. H.,** Evidence for a role of phosphatidylethanolamine as a modulator of membrane-membrane contact, *J. Membrane Biol.*, 48, 95, 1979.
47. **Evans, E. A. and Parsegian, V. A.,** Energetics of membrane deformation and adhesion in cell and vesicle aggregates, *Ann. N.Y. Acad. Sci.*, 416, 13, 1983.
48. **Boni, L. T., Hah, J. S., Hui, S. W., Mukherjee, P., Ho, J. T., and Jung, C. Y.,** Aggregation and fusion of unilamellar vesicles by poly(ethyleneglycol), *Biochim. Biophys. Acta*, 775, 409, 1984.
49. **Hui, S. W., Stewart, T. P., Boni, L. T., and Yeagle, P. L.,** Membrane fusion through point defects in bilayers, *Science*, 212, 921, 1981.
50. **Düzgünes, N., Wilschut, J., Fraley, R., and Pahadjopoulos, D.,** Studies on the mechanism of membrane fusion: role of head-group composition in calcium- and magnesium-induced fusion of mixed phospholipid vesicles, *Biochim. Biophys. Acta*, 642, 182, 1981.
51. **Düzgünes, N., Paiement, J., Freeman, K. B., Lopez, N. G., Wilschut, J., and Papahadjopoulos, D.,** Modulation of membrane fusion by ionotropic and thermotropic phase transitions, *Biochemistry*, 23, 3486, 1984.
52. **Sundler, R., Düzgünes, N., and Papahadjopoulos, D.,** Control of membrane fusion by phospholipid headgroups. II. The role of phosphatidylethanolamine in mixtures with phosphatidate and phosphatidylinositol., *Biochim. Biophys. Acta*, 649, 751, 1981.
53. **McIver, D. J. L.,** Control of membrane fusion by interfacial water: a model for the actions of divalent cations, *Physiol. Chem. Phys.*, 11, 289, 1979.
54. **Bentz, J. and Düzgünes, N.,** Fusogenic capacities of divalent cations and the effect of liposome size, *Biochemistry*, 24, 5436, 1985.
55. **Ekerdt, R. and Papahadjopoulos, D.,** Intermembrane contact affects calcium binding to phospholipid vesicles, *Proc. Natl. Acad. Sci. U.S.A.*, 79, 2273, 1982.
56. **Israelachvili, J. and Pashley, R.,** The hydrophobic interaction is long range, decaying exponentially with distance, *Nature (London)*, 300, 341, 1982.
57. **Ohki, S.,** A mechanism of divalent ion-induced phosphatidylserine membrane fusion, *Biochim. Biophys. Acta*, 689, 1, 1982.
58. **Nir, S., Wilschut, J., and Bentz, J.,** The rate of fusion of phospholipid vesicles and the role of bilayer curvature, *Biochim. Biophys. Acta*, 688, 275, 1982.
59. **Bentz, J., Nir, S., and Wilschut, J.,** Mass action kinetics of vesicle aggregation and fusion, *Colloid Surf.*, 6, 333, 1983.

60. **Knutton, S.,** Studies on membrane fusion. III. Fusion of erythrocytes with poly(ethyleneglycol), *J. Cell Sci.,* 36, 61, 1979.
61. **Bentz, J., Ellens, H., Lai, M.-Z., and Szoka, F. C.,** On the correlation between H$_{II}$ phase and the contact-induced destabilization of phosphatidylethanolamine-containing membranes, *Proc. Natl. Acad. Sci. U.S.A.,* 82, 5742, 1985.
62. **Ellens, H., Bentz, J., and Szoka, F. C.,** Destabilization of phosphatidylethanolamine liposomes at the hexagonal phase transition temperature, *Biochemistry,* 25, 285, 1986.
63. **Kwok, R. and Evans, E. A.,** Thermoelasticity of large lecithin bilayer vesicles, *Biophys. J.,* 35, 637, 1981.
64. **Rand, R. P., Kachar, B., and Reese, T. S.,** Dynamic morphology of calcium-induced interactions between phosphatidylserine vesicles, *Biophys. J.,* 47, 483, 1985.
65. **Zimmerberg, J., Cohen, F., and Finkelstein, A.,** Micromolar Ca^{2+} stimulates fusion of lipid vesicles with planar bilayers containing a calcium-binding protein, *Science,* 210, 906, 1980.
66. **Cohen, I. S., Zimmerberg, J., and Finkelstein, A.,** Fusion of phospholipid vesicles with planar phospholipid bilayer membranes. II. Incorporation of a vesicular membrane marker into the planar membrane. *J. Gen. Physiol.,* 75, 251, 1980.
67. **Akabas, M. H., Cohen, I. S., and Finkelstein, A.,** Separation of the osmotically-driven fusion event from vesicle-planar membrane attachment in a model system for exocytosis, *J. Cell Biol.,* 98, 1063, 1984.
68. **Miller, C., Arvan, P., Telford, J. N., and Racker, E.,** Ca^{2+}-induced fusion of proteoliposomes: dependence on transmembrane osmotic gradient, *J. Membr. Biol.,* 30, 271, 1976.
69. **Ohki, S.,** Effects of divalent cations, temperature, osmotic pressure gradient and vesicle curvature on phosphatidylserine vesicle fusion, *J. Membrane Biol.,* 77, 265, 1984.
70. **Ahkong, Q. F., Fisher, D., Tampion, W., and Lucy, J. A.,** Mechanisms of cell fusion, *Nature (London),* 253, 194, 1975.
71. **Chandler, D. E. and Heuser, J. E.,** Arrest of membrane fusion events in mast cells by quick freezing, *J. Cell Biol.,* 86, 666, 1980.
72. **Plattner, H.,** Membrane behavior during exocytosis, *Cell Biol. Int. Rep.,* 5, 435, 1981.
73. **Schmidt, W., Patzak, A., Lingg, G., Winkler, H., and Plattner, H.,** Membrane events in adrenal chromaffin cells during exocytosis: a freeze-etching analysis after rapid cryofixation, *Eur. J. Cell Biol.,* 32, 31, 1983.
74. **Gad, A. E. and Eytan, G. D.,** Chlorophylls as probes for membrane fusion. Polymyxin B-induced fusion of liposomes, *Biochim. Biophys. Acta,* 727, 170, 1983.
75. **Eytan, G. D. and Almary, T.,** Melittin-induced fusion of acidic liposomes, *FEBS Lett.,* 156, 29, 1983.
76. **Blumenthal, R., Henkart, M., and Steer, C. J.,** Clathrin-induced pH-dependent fusion of phosphatidylcholine vesicles, *J. Biol. Chem.,* 258, 3409, 1983.
77. **Schenkman, S., Araujo, P. S., Dijkman, R., Quina, F. H., and Chaimovich, H.,** Effects of temperature and lipid composition on the serum albumin-induced aggregation and fusion of small unilamellar vesicles, *Biochim. Biophys. Acta,* 649, 633, 1981.
78. **Garcia, L. A. M., Araujo, P. S., and Chaimovich, H.,** Fusion of small unilamellar vesicles induced by a serum albumin fragment of molecular weight 9000, *Biochim. Biophys. Acta,* 772, 231, 1984.
79. **Hong, K. and Vacquier, V. D.,** Fusion of liposomes induced by a cationic protein from the acrosome granule of abalone spermatozoa, *Biochemistry,* 25, 543, 1986.
80. **Doms, R. W., Helenius, A., and White, J.,** Membrane fusion activity of the influenza virus hemagglutinin. The low pH-induced conformational change, *J. Biol. Chem.,* 260, 2973, 1985.
81. **Stegmann, T., Booy, F. P., and Wilschut, J.,** Effects of low pH on influenza virus: activation and inactivation of the membrane fusion capacity of the hemagglutinin, *J. Biol. Chem.,* in press, 1988.
82. **Zimmerberg, J. and Whitaker, M.,** Irreversible swelling of secretory granules during exocytosis caused by calcium, *Nature (London),* 315, 581, 1985.

Chapter 14

THE ROLE OF THE ACTOMYOSIN SYSTEM IN SECRETION

James L. Daniel

TABLE OF CONTENTS

I. INTRODUCTION

All eukaryotic cells contain contractile proteins.[1,2] The discovery of the two major contractile proteins, actin and myosin, in such secretory cells as platelets, chromaffin cells, leukocytes, and the islets of Langerhans suggests that contractile proteins have a central role in energy-dependent exocytosis from these cells. In order to discuss the role of the contractile system of proteins in secretion, it is necessary to describe the biochemistry of the proteins of the contractile system.

II. MYOSIN AND MYOSIN-ASSOCIATED PROTEINS

A. Myosin

The structure of nonmuscle myosin has been extensively studied in only a few cells where its structure is similar to that of muscle myosin. It is probably safe to assume that myosin structure is reasonably constant in mammalian cells. Myosin is a hexamer with a native molecular weight of about 480 kdaltons.[3] Each molecule of myosin contains two heavy chains of about 200 kdaltons and four smaller subunits that have been called light chains. The light chains have molecular weights in the range of 16 to 21 kdaltons depending on cell type. Skeletal muscle myosin contains two 18.5 kdalton light chains and either two 21-kdalton light chains or two-17 kdalton light chains.[4-6] Smooth muscle myosin contains 20-kdalton and 16-kdalton light chains.[7] Nonmuscle myosins have been shown to be similar to smooth muscle myosin.[8,9] The heavy chain of myosin is a polar structure. The C-terminal half of the heavy chain is almost totally α-helical and the N-terminal region of the myosin heavy chain forms a globular "head" region. The α-helical regions from the two heavy chains interact to form a 140-nM long coiled coil which has been termed the myosin rod.

Proteolytic cleavage has been very useful in elucidating the structure of myosin.[3] Under the proper conditions, myosin can be cleaved into a region known as light meromyosin (LMM) that contains most of the rod and heavy meromyosin (HMM) that contains the two heads joined by a small section of rod. HMM can be further cleaved to yield two separated heads, known as subfragment one (S-1) and the small rod region, known as subfragment two (S-2). Since the light chains are found associated with S-1, the light chains can be assigned to the head region of myosin. This region also contains both the actin-binding site and the ATP hydrolysis site of myosin. Removal of both light chains is associated with loss of enzymatic activity, implying either a direct role of the light chains in enzymatic site or that the light chains stabilize the three-dimensional structure of the head.[10]

At physiological salt concentrations, skeletal myosin molecules aggregate into 1600 by 15-nm filaments which are very similar in dimension to the thick filaments of muscle.[11] The tail region of the myosin molecule forms the core of the filament. LMM can form filaments similar in size to those of intact myosin. The S-2 fragment is soluble in physiological buffers and it is thought that this region is loosely associated with the filament backbone and allows the heads to extend to the actin filament during actin-myosin interaction. Each filament of skeletal muscle myosin contains about 300 myosin molecules. The filaments of platelet myosin are formed in the same manner as those of skeletal muscle. However, filaments formed from platelet myosin are much smaller than skeletal muscle myosin filaments. Purified platelet myosin forms 325 by 11 nm filaments with each filament containing 28 to 30 myosin molecules.[12] These small filaments would be very difficult to detect in transmission electron micrographs of nonmuscle cells possibly explaining the fact that myosin filaments are seldom observed in these cells.

The content of myosin in nonmuscle cells can be somewhat variable. Estimates of myosin content based both on ATPase measurements and gel densitometry indicate that nonmuscle cells contain 0.3 to 1.5% of their protein as myosin.[13] This can be compared to skeletal muscle cells in which myosin comprises 35% of total protein. Platelets are particularly rich in myosin with estimates that 2 to 5% of total protein is myosin.[14]

B. Myosin Light Chain Kinase

Myosin light chain kinase uses one molecule of ATP to phosphorylate a specific serine on myosin. Platelet myosin light chain kinase was the first nonmuscle kinase isolated, using a calmodulin affinity column.[15,16] The kinase was not active unless Ca^{2+} was added and the Ca^{2+}-binding protein calmodulin was present. Hathaway and Adelstein[17] also isolated the kinase from platelets and found that it had a molecular weight of 105 kdaltons and a specific activity of 3.1 μmol of Pi transferred to light chain per milligram per minute. It has K_m's for ATP and light chain of 121 and 18 μM, respectively. The kinase did not phosphorylate phosvitin, casein, histone IIIS, glycogen phosphorylase b, or rabbit skeletal muscle troponin indicating the specificity of the kinase for the regulatory light chain. Originally, platelet myosin light chain kinase was isolated as an 83-kdalton protein which was not Ca^{2+}-regulated.[15] This protein probably represents a proteolytic fragment of the intact kinase.[18]

Calmodulin-dependent myosin light chain kinases have been isolated from brain[19] and baby hamster kidney (BHK-21) cells.[16] Macrophages contain kinase but the regulation of this kinase by Ca^{2+} has not been studied.[20] Myoblast kinase from rats was not Ca^{2+}-dependent but may represent a proteolytic fragment of a calmodulin-regulated kinase.[21]

C. Phosphatase

Much less information is available concerning the phosphatase that dephosphorylates nonmuscle myosins. Adelstein and co-workers[22] have isolated a crude fraction from platelets that contains phosphatase activity. The phosphatase dephosphorylated ^{32}P-labeled myosin and isolated light chains.

The phosphatase of smooth muscle has been studied in much greater detail. The phosphatase was purified using affinity chromatography on a column of thiophosphorylated light chains coupled to Sepharose 4B as the final step.[23] The phosphatase had a molecular weight of 165 kdaltons and contained three polypeptides (60, 55, and 38 kdalton) in a 1:1:1 stoichiometry.[24] This enzyme did not require Mg^{2+} for activity and does not dephosphorylate intact myosin, but does dephosphorylate isolated light chains or myosin light chain kinase. However, the 38-kdalton catalytic subunit will dephosphorylate isolated light chains, intact myosin, or myosin light chain kinase. This suggests that the phosphatase might be regulated by association of the catalytic subunit with the two other subunits.

D. Calmodulin

Calmodulin is a very acidic protein of 17 kdaltons with four binding sites for calcium.[25] This protein was originally recognized as a calcium-dependent regulator of brain phosphodiesterase. Later, it was realized that calmodulin could regulate, in a calcium-dependent fashion, other enzymes including brain adenylate cyclase, gizzard myosin light chain kinase, and erythrocyte membrane ATPase. The amino acid sequence of calmodulin is similar to that of muscle troponin-C and many regions are homologous to each other suggesting that both proteins were formed by multiple gene duplication. Binding of calcium to calmodulin appears to increase the α-helix content of the protein at the expense of random coil. Activation of different enzyme systems may require a different number of bound calciums. Blumenthal and Stull[26] studied the activation of skeletal muscle myosin kinase by varying both the calmodulin concentration and the free calcium concentration. They concluded that the kinase forms a 1:1 complex with calmodulin and that binding of calcium at all four sites is required. Investigators from several laboratories have used a variety of techniques to demonstrate that maximal activity of smooth muscle myosin kinase requires a 1:1 complex of kinase with calmodulin.[27,28]

E. cAMP-Dependent Protein Kinase

cAMP-dependent protein kinase does not interact directly with myosin, but phosphorylates myosin light chain kinase. Phosphorylation of the kinase decreases the activity of the enzyme primarily by lowering the affinity of the myosin light chain kinase for calmodulin 5- to 20-fold, although some effect on V_{max} was also found for the platelet enzyme.[29,30] In bovine aortic actomyosin,[31] cAMP-dependent phosphorylation of myosin kinase was found to increase the level of free Ca^{2+} needed for half-maximal activation from 1 to 2.5 μM. Adelstein[29-31] has proposed that phosphorylation of the myosin light chain kinase is a major mechanism by which agents which elevate cAMP, inhibit platelet and other nonmuscle cell functions. However, some investigators[32] have questioned whether this mechanism could operate in intact cells since excess calmodulin is present.

F. Protein Kinase C

Protein kinase C is a protein kinase first isolated from brain by Nishizuka.[33] Protein kinase C is activated by Ca^{2+} but does not require calmodulin, but rather requires phospholipid and diacylglycerol for full activation. This kinase can be activated by phorbol esters in place of diacylglycerol. Myosin is phosphorylated by protein kinase C at a different site from that phosphorylated by myosin light chain kinase.[34,35]

III. ACTIN AND ASSOCIATED PROTEINS

A. Actin

Actin is a major protein of the cytosol of nonmuscle cells. Platelets are the nonmuscle cell that contains the greatest amount of actin; it is estimated that 20 to 30% of total platelet protein is actin.[36] In other nonmuscle cells, actin is about 5 to 10% of total protein.[37] The importance of actin to the functioning of eukaryotic life is perhaps best illustrated by the fact that actin is ubiquitous and its amino acid sequence is highly conserved. It is extremely difficult to raise antibodies against actin unless it has been denatured by the detergent sodium dodecyl sulfate (SDS).[38] This low antigenicity probably is a result of the high conservation of the actin sequence, although there are small sequence differences.[39]

The mobility of nonmuscle actins is identical to skeletal muscle actin on SDS polyacrylamide gel electrophoresis (PAGE). This indicates that it has about the same molecular weight, 42 kdaltons. However, more than one form of actin is present in mammalian organisms for when a mixture of different actins is electrophoresed on isoelectric focusing gels, at least three bands have been obtained.[40] The most acidic actin band has been called α-actin and is the major component of muscle actin. The other actins are named β- and γ-actin and are found in nonmuscle tissues. α-Actin has been completely sequenced. Peptides from the different forms of actin show that the actins are all products of different genes and that β- and γ-actin are more closely related to each other than to α-actin.[41] The β-γ ratio can vary with both cell type and species. The importance of these two forms of actin in cells is unclear and the two forms copolymerize in in vitro experiments.[42] In platelets β- and γ-actin are distributed evenly between the monomeric, filamentous, and membrane pools of actin.[43]

Actin has been crystallized as a complex with DNAase I.[44] The actin monomer has dimensions of $3.7 \times 4.0 \times 6.7$ nm and consists of a large and a small lobule separated by a cleft. Actin has one binding site for nucleotide-divalent metal complex, usually calcium.[45] This nucleotide may be in the cleft between the two lobules. As a monomer (monomeric actin is called G-actin and filamentous actin is called F-actin), the binding of ATP is preferred over ADP and since ATP is at ten times higher concentration than ADP in the cytosol of cells,[46] it is likely that almost all G-actin in the cytosol contains ATP rather than ADP. The bound nucleotide serves to stabilize the G-actin structure since monomeric actin rapidly denatures if its bound nucleotide is lost.

At low salt concentrations and low concentrations of divalent cation, actin will remain as monomers.[42,45] Raising the KCl concentration to 50 mM and/or the Mg^{2+} concentration to 2 mM will cause the monomers to polymerize into a long strand of F-actin. When actin polymerizes, its bound ATP is hydrolyzed to ADP; thus, the bound nucleotide of F-actin is ADP. The binding of this ADP is very tight and perhaps it would be totally inexchangable if it were not for exchange occurring at the ends (see below). The F-actin filament contains two strands of actin monomers arranged in two intertwining right-handed helices with a pitch of about 2×36 nm and 6 to 7 nm in diameter;[47] the longest dimension of the monomer is aligned perpendicular to the filament axis. The actin filament has polarity as evidenced by the binding of subfragment 1 (S-1; the myosin head produced by proteolysis of myosin [see below]). In the electron microscope, actin decorated with S-1 appears as a series of arrowheads; all the arrowheads of one filament point in the same direction. The end to which the S-1 arrowhead points has been called the "pointed" end and the other end, the "barbed" end. In muscle, the actin filament is the backbone for the thin filaments of the I-band and is about 1 μm long.

If only G-actin monomers are present when the salt concentration is increased, the polymerization initially exhibits a lag phase prior to the onset of polymerization.[48] This lag is thought to reflect the first phase of polymerization and nucleation, in which two or three actin monomers must aggregate in the specific conformation required for further polymerization. Once these nuclei are formed, more rapid polymerization occurs through addition of monomers to these small polymers. If a sonicate of F-actin is added to G-actin with KCl and Mg^{2+}, polymerization proceeds rapidly with no evidence of a lag.

All the actin monomers do not become incorporated into the polymer. When polymerization has reduced the free monomer concentration to a certain concentration called the critical concentration, net polymerization stops. At the critical concentration, polymerization continues but the monomers disassociate from the polymer ends at the same rate as monomers are added. If the initial monomer concentration (i.e., when the salt concentration is raised) is below the critical concentration, no polymerization occurs; dilution of F-actin below this concentration will cause all filaments to depolymerize. While ADP will support polymerization, ATP causes polymerization at a lower critical monomer concentration than ADP.

The polymer does not grow at the same rate from both ends; growth is more rapid from the barbed end. This occurs primarily because the association rate constant at the barbed end is four times higher than at the pointed end.[49] As a result, the critical concentration at the two ends is different; the measured critical concentration in an actin solution is intermediate between the values at either end. When net polymerization has stopped, the monomer concentration is below the critical concentration for the pointed end but above the critical concentration of the barbed end. In this steady-state condition, monomers are being continuously removed from the pointed end and are being added to the barbed end. This process has been called "treadmilling" because a given monomer, if it could be watched, would move from the barbed toward the pointed end, then off; just as if it were moving along a treadmill. When the monomers leave, ADP is immediately exchanged for ATP, while monomers added to the filament hydrolyze their bound ATP. Thus, energy is being constantly consumed. This energy is required to keep the treadmill moving; otherwise the treadmill would violate the laws of thermodynamics. Nonhydrolyzable ATP analogs allow actin to polymerize with the same critical concentration as occurs in the presence of ATP, but of course the analogs will not support treadmilling. It is not clear exactly what importance this process plays in the homeostasis of the living cell, but it is possible that the mechanism could be used to move materials within the cytoplasm.

Nonmuscle actin, at physiological salt and temperature, behaves very much like skeletal muscle actin. The free monomer concentration is about 0.03 to 0.06 mg/mℓ. However, at low temperature or at higher salt concentrations more actin is depolymerized; i.e., nonmuscle actin has a higher critical concentration than muscle actin.[50] If 10% of a cell's protein is actin, then the concentration of cytoplasmic actin is at least 7 mg/mℓ.[43] With a critical concentration of about 0.5 mg/mℓ, greater than 95% of cell's actin should be polymerized. However, several lines of evidence suggest that this may not be the case and that a substantial portion of nonmuscle cell's actin is depolymerized. This evidence will be discussed below.

B. Tropomyosin

Muscle tropomyosin is a rod-shaped dimer with a length of 39 to 40 nm and has a molecular weight of about 33 kdaltons. It contains over 95% α-helix and is probably a coiled-coil α-helical rod. Platelet tropomyosin, the best characterized nonmuscle tropomysin, is slightly smaller with a molecular weight of 28 kdaltons and rod length of 34 to 35 nm.[51] The reduced size is due to deletion of 37 amino acids from the N-terminal end of the protein. Tropomyosin is found in the grooves of the actin thin filament. The binding of muscle tropomyosin to actin is cooperative, probably due to tail-to-tail interactions of tropomyosin molecules. One molecule of muscle tropomyosin is associated with seven molecules of actin, i.e., it is long enough to bind to seven protomers in the F-actin filament. Platelet tropomyosin binds to only six actins because of its reduced length. The association of platelet tropomyosin with actin is weaker than for muscle tropomyosin, probably due to the smaller size of platelet tropomyosin and a resulting lack of tail-tail interaction.[52] There is no apparent difference in the ability of platelet and muscle actins to interact with tropomyosin. There is evidence for an Mg^{2+} regulation of the interaction of nonmuscle tropomyosin and F-actin.[42]

Lazarides[53] has proposed based on immunofluorescence that not all the actin filaments of nonmuscle cells contain tropomyosin. Filaments in larger, more rigid actin bundles contain tropomyosin, while filaments in more dynamic regions of the cell have no tropomyosin.

This suggests that tropomyosin has a stabilizing function on actin microfilaments (microfilaments is the term used for 6 nM actin filaments in nonmuscle cells). While a 1:6 ratio of tropomyosin to actin monomer would be required for platelet actin to be saturated with tropomyosin, in platelets the ratio is 1:14. Based on the fact that tropomyosin binding to actin is cooperative, Harris[43] has proposed that when platelet actin is fully polymerized, two classes of actin filaments might exist in platelets. One class of actin filaments would contain tropomyosin and the other class would not. Harris has suggested that the two types of filaments might have different subcellular localizations and presumably different functions. Tropomyosin enhances the ATPase of platelet actomyosin possibly by stiffening the actin filament and, thus, facilitating interaction with myosin filaments. In addition, tropomyosin influences the interaction between phosphorylated myosin and actin (see below) and inhibits fragmentation of actin by actin severing proteins.

C. Profilin

Profilin is a 15.2-kdalton protein that was originally isolated from spleen and thymus[54] and later found in platelets,[36,55] lymphocytes,[56] brain,[56] thyroid,[57] and macrophages.[58] It is a basic protein that forms 1:1 reversible complex with an actin monomer. The actin-profilin complex is called profilactin. Actin that is complexed with profilin is not able to polymerize. Addition of 0.3 to 1 mol of profilin per mole actin slows the rate of polymerization. However, the same steady state concentration of F-actin is obtained if profilin is added to F-actin or if the same amount of G-actin is polymerized in the presence of profilin.[42] Because it lowers the free monomer concentration profilin inhibits nucleation, a step that has a high dependence on the G-actin concentration. Profilin also slows actin polymerization, but to a lesser degree than it inhibits nucleation. Acanthamoeba profilin stimulates the exchange rate of actin-ATP with free ATP.[59] Lassing and Linberg have made the interesting observation that phosphatylinositol 4,5-bisphosphate can dissociate profilactin.[60]

Profilin shows no preference for β- or γ-actin.[36,56] As mentioned above, not all the actin of nonmuscle cells is polymerized. This fraction of actin which does not polymerize has been estimated to be 30 to 70% of the total actin, much more than can be explained by the critical concentration.[36,54] There is enough profilin in platelets to bind all the unpolymerized actin. However, since this reaction is a reversible equilibrium, Korn[42] has calculated that if profilin were twice the steady state concentration of actin monomer, only one third of the actin would be depolymerized as a result of profilactin formation. Thus, the presence of profilactin in nonmuscle cells could only partially explain the unpolymerized actin of platelets; other protein interactions are probably also involved (see below).

D. DNAase I

Pancreatic DNAase I forms a tight 1:1 complex with G-actin.[61] The K_d of this complex is 2 nM. Some evidence suggests that DNAase I can bind directly to F-actin but it not clear whether DNAase I can directly depolymerize actin or that it causes depolymerization in the same manner as profilin.[61,62] The enzymatic activity of DNAase I is fully inhibited when it is bound to G-actin.[63,64] The inhibition of DNAase activity has been used to determine the percentage of actin that is present as monomer in an unknown sample.[65] This method has been applied to several different cells but suffers from the disadvantage that the cells must be broken prior to assay for inhibition of DNAase activity.

E. Gelsolin

Gelsolin, a 91-kdalton protein first isolated from lung macrophages[65] was later isolated from many cells including muscle.[37] Gelsolin binds to the fast growing barbed end of actin. Gelsolin promotes nucleation of actin filaments presumably by binding to and stabilizing actin nucleation complexes. Thus, gelsolin, if added to a polymerizing solution of actin monomers, would cause the formation of both many more and shorter filaments than would form in its absence. Gelsolin added to a solution of preformed actin filaments causes a redistribution to a larger number of shorter filaments. However, the rate of this redistribution is too fast to be simply explained by the end blocking activity of gelsolin alone. This rapid change in filament length has been attributed to a severing action by which gelsolin binds to the middle of an actin filament and breaks actin-actin bonds. These activities combine to reduce the viscosity of actin solutions and thus can control the viscosity of the cytosol of secretory cells.

The concentration of Ca^{2+} appears to be an important regulator of the expression of the three effects of gelsolin on actin, severing, nucleation, and end blocking.[66-68] Severing requires the highest level of Ca^{2+}, above 1 μM. Binding to the barbed end will occur below 1 μM Ca^{2+} but is considerably enhanced at higher Ca^{2+}. Gelsolin forms a complex with two G-actins in the presence of micromolar Ca^{2+} concentrations and this complex containing at least one tightly bound Ca^{2+} can be considered to be an actin nucleation site. Blood plasma contains a slightly larger form of gelsolin (93 kdaltons) that has been called brevin.[69]

F. Acumentin and β-Actinin

Acumentin is a 65-kdalton monomeric protein that has been found in macrophages[70] and granulocytes[71] in high concentration (6% of cytoplasmic extracts). The role of acumentin in regulation of actin polymerization is less clear than that of gelsolin. It binds weakly to the slow polymerizing pointed end of F-actin and can cause nucleation. It has little effect on the critical concentration of F-actin solutions.

β-Actinin first found in muscle is a 80-kdalton monomer in rat kidney cells.[72] β-Actinin enhances nucleation and shortens actin filaments, an activity that is enhanced in the presence of millimolar $MgCl_2$. It may bind to the pointed end of the filament.

G. Actin-Binding Protein

Actin-binding protein (ABP) is a high molecular weight dimer with a subunit molecular weight of about 260 kdaltons that was originally purified from rabbit lung macrophages[73] and later from platelets[74,75] and BHK-21 cells.[76] ABP is a major cellular protein accounting for 1 to 2% of leukocyte protein and 7% platelet protein. ABP is highly elongated with a measured axial ratio of 17:1.[77] In the electron microscope, the isolated dimer appears as a 162-nm flexible rod with a 3-nm diameter. This protein is very similar in its properties to a smooth muscle protein of slightly lower molecular weight called filamin.[78] ABP binds actin with a stoichiometry 1:14, i.e., about every 40 nm along the filament at high ABP concentration.[73] The main activity associated with actin-binding protein is crosslinking actin filaments to form a gel.[79,80] The F-actin gels formed in the presence of low concentrations of ABP are isotropic, i.e., they form a network or mesh rather than bundles. Actin filaments crosslinked by ABP form arrays in which the filaments are perpendicular to each other; a molecule with flexibility or a right-angle hinge would be required to construct such crossbridges. ABP is the only protein known to localize at filament crossover points. At higher concentrations, ABP may induce bundle formation.

Platelet ABP has been shown to be a phosphoprotein.[81] ABP protein from resting platelets contains 2 mol of phosphate per mole. Removal of 1 mol of phosphate removed the crosslinking activity of ABP, and dephosphorylated ABP no longer binds to an F-actin affinity column. Changes in the phosphorylation state of ABP during platelet activation have been reported.

H. Caldesmon

Caldesmon is a dimeric protein of 120-kdalton subunits from chicken gizzard that both crosslinks actin filaments and binds to calmodulin.[82] The binding to calmodulin inhibits crosslinking. Caldesmon has been found in brain where it has a range of molecular weights.[83] A 79-kdalton form of caldesmon (but no higher molecular weight forms) has been associated with platelet actin.[84] Adrenal chromaffin cells contain a 70-kdalton form of caldesmon.[85] Chromaffin-cell caldesmon appears to be associated with chromaffin granules and its interaction with actin is inhibited by Ca^{2+}.

I. α-Actinin

α-Actinin is considered a bundling protein that stabilizes the side-by-side association of actin into aggregates. α-Actinin is a major protein of the Z-line of skeletal muscle where it serves as an anchor for the I-band.[1] Muscle α-actinin is a dimer with a dimension of about 2 to 4 by 30 to 40 nm, contains about 74% α-helix and consists of two 105-kdalton subunits.[86,87] In vitro, α-actinin interacts with actin to form gels. This interaction is inhibited by tropomyosin. The binding of α-actinin appears to occur at the ends of the rods and saturates at a 1:10 ratio to actin.[86] At low ratios of α-actinin to actin, crosslinking occurs,[88] at higher ratios, bundles can form and α-actinin is found at 30-nm intervals along the bundles.[86,87] ABP and α-actinin can cooperate to form parallel arrays of F-actin filaments.[75] In addition, α-actinin at a molar ratio of 1:200 can dissociate profilactin and promote actin polymerization.[89] Of course this effect may be due to the crosslinking properties of α-actinin.

α-Actinin has been shown to be in a wide variety of cells, including various cell culture lines, platelets, brain cells, and macrophages. Platelet α-actinin has been isolated and characterized and found to contain two subunits with slightly different molecular weights (100 and 102 kdaltons) but is otherwise similar to muscle α-actinin.[90] The interaction between nonmuscle α-actinins and actin is Ca^{2+} sensitive.[91] Actin gels are formed at Ca^{2+} concentrations below 10^{-7} M and inhibited at Ca^{2+} above this concentration. Immunofluorescence studies show that in some nonmuscle cells, α-actinin is found near the membrane.[53,92] In cultured cells α-actinin is found near focal adhesion plaques but not directly linked to the membrane.[92,93] α-Actinin can be removed from membranes under mild conditions where actin remains attached, indicating that α-actinin does not link actin directly to the membrane components.[94]

J. Vinculin and Talin

Vinculin is a roughly globular, 130-kdalton, actin cross-linking protein first purified from smooth muscle.[92,95] It is found at the end of stress fibers (bundles of actin filaments found in cultured cells) near the leading edges of cells and in or near the adhesion plaques of fibroblasts. It is easily extracted from cells without detergents and thus, does not appear to be an integral membrane protein; but it may, however, participate in anchoring actin filaments to membranes.[95,96] Vinculin, in contrast to other actin gelation factors, reduces the low shear viscosity of F-actin solutions. Some investigators[95] have suggested that this reduction in viscosity is due to a bundling of actin filaments into a tightly packed, almost paracrystalline fiber, while others have disputed this claim and have proposed that vinculin binds to filament ends and thus causes shorter filaments to be formed.[96,97] The interaction of muscle vinculin with actin is not Ca^{2+} sensitive, but HeLa cell vinculin decreases the viscosity of F-actin solution only when Ca^{2+} concentration is above 1 μM.[96] Platelets contain vinculin of similar properties.[98] Vinculin is one of the substrates of the tyrosine protein kinase pp60[src].[98] It has been suggested but not proven that phosphorylation of vinculin at a tyrosine might alter its interaction with actin at adhesion plaques and may in part account for the morphologic change seen with transformation.

Talin is a 215-kdalton smooth muscle protein that is found in association with vinculin at adhesion plaques that binds to vinculin but not actin.[100,101] It may help couple actin filaments to integral membrane proteins. A talin-like 235-kdalton protein is found in platelets where it is called P235.[102,103] Interestingly, this protein is cleaved by a calcium-dependent protease when the cells are activated. Both talin and P235 may be phosphoproteins whose phosphorylation is catalyzed by protein kinase C.

K. Spectrin and Ankyrin

Spectrin and ankyrin are found closely associated with the membrane of red blood cells.[104] Spectrin is a heterodimer with subunits of 220 and 240 kdaltons. Spectrin is a relatively rigid rod which associates into a 200-nm tetramer with binding sites for actin at either end. Much more spectrin is required to cause actin gelation than ABP, a fact attributed to the inflexibility of spectrin.[105] Spectrin-actin complexes seem to be two-dimensional sheets in agreement with the proposed role of this complex as a supporting "cytoskeleton" for the red cell membrane. Ankyrin forms a link between the spectrin actin-complex and an integral red cell membrane glycoprotein.[104] Another protein, band 4.1, increases the affinity of spectrin for actin by 10- to 100-fold. Spectrin can also bind calmodulin with no known function yet ascribed to this interaction.

Spectrin-related proteins have been detected in platelets,[106] brain,[107] and intestinal epithelial cells.[108] The brain and epithelial spectrin have been named fodrin or TW260/240. These proteins are very similar in physical and biochemical properties to erythrocyte spectrin. There is also immunologic evidence for the presence of ankyrin in platelets.[109] These proteins may be important in the interactions of actin with membranes.

IV. CYTOCHALASINS AND PHALLOIDIN

Cytochalasins are not proteins but a family of low molecular weight fungal metabolites that have proved useful in the study of the nonmuscle contractile system.[1,42] Cytochalasins were shown to inhibit cell locomotion, phagocytosis, and cell division but not contraction of skeletal muscle.[1] They have been widely used in secretory cells to study the role of contractile proteins in secretion. Lin and co-workers were the first to define the site of action of cytochalasins when they showed that high affinity binding sites for [^{3}H]-cytochalasin B were present in red cell ghosts.[97] These sites could serve as actin nucleation sites. It was shown that these sites are actin polymers. The binding of cytochalasins to F-actin is very tight with a K_d of 10^{-8} M or below but does not bind to actin monomers. Studies using the electron microscope have shown that cytochalasin B blocks elongation at the rapidly growing barbed end of the filament. The action of cytochalasins is similar to that of gelsolin in that net depolymerization of long filaments and formation of polymerization nuclei lead to shorter filaments. The use of these compounds in the study of secretion will be discussed below.

Phalloidin is a bicyclic peptide from toadstools that binds to and stabilizes actin filaments.[42] In humans, its site of toxicity is the liver parenchymal cells where it causes a formation of stable actin filaments. It binds with 1:1 stoichiometry to the protomers of the F-actin filament. It accelerates polymerization, lowers the critical concentration, and prevents polymerization of formed filaments. This toxin has not been as useful as cytochalasins primarily because of its limited ability to enter cells but its dramatic effect in the liver serves to emphasize the importance of a dynamic actin polymerization-depolymerization system to healthy function of cells.

V. REGULATION OF NONMUSCLE ACTOMYOSIN

It is well established that skeletal myosin is regulated by Ca^{2+} which binds to the troponin-tropomyosin system.[1] This system is called actin-linked because troponin and tropomyosin are found on the actin filaments. A feature of this system is that purified skeletal actomyosin exhibits high activity (ATPase) in the absence of Ca^{2+}. Thus, troponin provides regulation by inhibiting actomyosin when Ca^{2+} is absent. In nonmuscle cells, the evidence for an actin-linked system is rather poor. The nonmuscle contractile system is activated by phosphorylation of myosin light chain by myosin light chain kinase which is regulated via calmodulin.[110] In contrast to the skeletal muscle system, purified nonmuscle actin and myosin exhibit a very low ATPase activity and phosphorylation is required to activate the system. While some studies have been made on other nonmuscle cells most of the information on the role of phosphorylation in relating nonmuscle contractility comes from studies on platelets or smooth muscle whose contractile activity is regulated in a very similar manner to platelet actomyosin.

A. The Effect of Phosphorylation on the Activity of Platelet Actomyosin
1. Effects on ATPase

The contractility of actomyosin systems is difficult to measure in a test tube. Because the hydrolysis of ATP provides the energy for muscle contraction, biochemists have used the actin-activated ATPase as an indicator of the contractility of both muscle and platelet actomyosin. Adelstein and Conti showed that when the ATPase activity of phosphorylated and control platelet myosin was measured, phosphorylation had no effect on the ATPase of myosin alone.[111] However, the activity of the actin-activated ATPase of the platelet myosin was enhanced about sixfold by phosphorylation. The ATPase of smooth muscle myosin has also been shown to be enhanced by phosphorylation.[112,113] Sellers et al.[114] used smooth muscle phosphatase I to totally dephosphorylate both turkey gizzard and platelet myosin. They also measured the phosphorylation state of the myosin using alkaline-urea PAGE. Smooth muscle myosin showed a 12-fold increase in the actin-activated ATPase after phosphorylation; while platelet myosin gave an 18-fold increase. Furthermore, the changes in the actin-activated ATPase activity paralleled the phosphorylation state through several cycles of phosphorylation-dephosphorylation.

2. Effect on the Contractility of an Actomyosin Thread

Lebowitz and Cooke[115] used a unique experimental model for contractility to add additional support to the theory that the interaction of actin and myosin was regulated by phosphorylation. In this model, actomyosin threads are mounted on a tensiometer in order to measure the isometric tension produced. The phosphorylation state of the samples was measured by alkaline-urea PAGE. It was found that the amount of isometric tension was directly proportional to the phosphorylation state of the platelet myosin. These studies indicate that the phosphorylation state of platelet myosin determines the contractility of platelet contractile proteins.

3. Effect of Contractility of Intact Platelets

The aspect of platelet function with the clearest analogy to muscle contraction is the phenomenon known as clot retraction. In order to study the role of platelet contractile proteins in the process of clot retracction, a good method of quantitation of the tension generated by the platelets is necessary. Salganicoff and co-workers[116,117] have developed a model system in which the force generated by contracting platelets can be measured with minimal interference from fibrin strands. Platelets are spun from cold plasma onto a nylon mesh. The thrombin, generated during preparation, activates the platelets which form a tight aggregate. The platelet mass and associated nylon can be cut into strips of tissue similar to smooth muscle strips and hung in an organ bath from a transducer. This system has been used to investigate the role of myosin phosphorylation in the force generation by platelet aggregates. Daniel et al.[118] and Bromberg et al.[119] found that when the preparation was treated with varying concentrations of either PGE$_1$ or PGI$_2$, the steady-state level of tension showed an inverse dose dependence on the amount of added PGE$_1$. The level of tension at each concentration of PGE$_1$ was correlated to the level of myosin light chain phosphorylation.

Kinetic measurements made on the rate of PGE_1-induced tension decrease and dephosphorylation of myosin light chain show that both decreased in parallel. After washing out the PGE_1, recovery of phosphorylation preceded the recovery of tension production. Bromberg et al.[119] also studied agonist-induced changes in tension and phosphorylation of myosin. In all cases the steady-state level of tension was correlated to the corresponding level of myosin phosphorylation.

4. Effect on filament formation

Smooth muscle, thymus, and platelet myosin filaments in 0.1 M KCl will disassemble if ATP is added.[120,121] Filaments formed from phosphorylated myosin are not disassembled by ATP. Skeletal myosin filaments were unaffected by increasing the ATP concentration. When smooth muscle myosin light chain kinase and Ca^{2+} were added to the ATP disassembled thymus myosin, the filaments reformed. The filament formation correlated with an increase in the phosphorylation state of the myosin light chain. It is not clear at present whether this mechanism can regulate filament formation in intact cells.[122,123]

VI. THE ROLE OF THE CONTRACTILE PROTEINS IN SECRETION

In this section, we will examine the evidence for a role of contractile proteins in the secretory mechanism of a few examples of secretory cells.

A. Insulin Secretion from Pancreatic B Cells

Secretion of insulin by the pancreatic B cells can be considered to be a reasonably continuous process with up to 10% of the 13,000 granules of a cell being secreted in each hour.[124] This contrasts with secretion by a cell like the blood platelet which does not receive continual stimulation and secretes all of its granules in less than a minute. Thus, it is not surprising that some details of the secretory mechanism would differ among different types of secretory cells. Secretion can be considered to occur in two phases: (1) translocation of granules to the plasma membrane and (2) fusion of the granule and plasma membranes (exocytosis).

1. Microtubules

Microtubules or 25-nm filaments have been implicated in insulin secretion due to the effect of drugs known to interfere with microtubule polymerization. Microtubule polymerization is very similar to actin polymerization and two types of drugs that interfere with microtubule polymerization are available (see Reference 1 for a discussion of microtubule biochemistry). Colchicine and nocodazole,[125,126] which cause depolymerization of microtubules by slightly different mechanisms, inhibit glucose-induced insulin secretion. Vinblastine,[127] which causes paracrystals of microtubules to form and taxol[126] which stabilizes microtubules both inhibit insulin secretion. These results suggest that a dynamic turnover of microtubules is important in insulin secretion. Furthermore, since a variety of microtubule-specific drugs produce the same effect, it is not likely that the drugs inhibit secretion through a nonspecific effect. In addition to these studies, a few studies indicate that microtubules associate with insulin-containing granules in the presence of microtubule-associated proteins (MAPs)[128] or in 1 mM $CaCl_2$.[129]

2. Contractile Proteins

The contractile proteins of the B cell are less well characterized than those of other cells. Actin was found in B cells in 1974 and accounts for 1 to 2% of total cell protein.[130] Microfilament bundles are visualized near the plasma membrane in a structure called the "cell web".[131] Polymerized actin as assayed by a DNAase I inhibition assay is increased from 30 to 50% of total actin upon stimulation of resting cells with glucose.[132,133] Myosin has been shown to exist as heavy and light chains and constitutes 0.5 to 1% of total cell protein.[134] Myosin light chain kinase activity has been demonstrated in extracts from both islet and insulin-secreting tumor cells.[135,136] Actin-associated proteins have not been fully characterized but two calcium-dependent factors of 200 and 40 kdaltons that can prevent actin polymerization have been demonstrated in islet cytosol preparations.[137]

The potential for interaction of actin with insulin-containing granules has been demonstrated. When isolated granules are centrifuged in the presence of F-actin or actomyosin, their sedimentation is retarded.[138] This action was enhanced by ATP. However, a nonspecific interaction cannot be ruled out. The possibility of a cooperative interaction between microtubules and microfilaments has been raised. F-actin and microtubules can interact in the presence of (MAPS).[139,140] Phosphorylation of MAPS inhibits the interaction.

In spite of the demonstration of myosin light chain kinase activity in B cells, the evidence for a phosphorylation of myosin light chain in response to glucose stimulation is not convincing at present.[141] Cytochalasin B has been shown to enhance secretion.[142] When granule movement within intact cells is measured by cinemicrography, cytochalasin B was shown to stimulate or not affect granule movement.[142,143]

B. Secretion from Adrenal Medullary Chromaffin Cells

1. Microtubules

Microtubules have been clearly demonstrated in chromaffin cells by immunofluorescence.[144,145] Studies of the cells at various stages in cell culture show that changes in cell shape are associated with redistribution of microtubules. Addition of colchicine caused the microtubule network to disappear and the cultured cells to round up. The distribution of granules in these cells is closely associated with the distribution of microtubules indicating that they are bound to microtubules. High affinity binding sites for tubulin have been demonstrated on chromaffin granule membrane.[144,146] Tubulin binding sites were also demonstrated on both plasma membrane and mitochondria. Secretion of catecholamines induced by K^+ is not blocked by either colchicine or vinblastine.[147]

2. Contractile Proteins

Contractile proteins of the chromaffin cells are better characterized than those of B cells. Myosin has been purified and has the same subunit composition and morphologic appearance as other nonmuscle myosins.[9] Its amino acid composition is comparable to that of other myosins. It forms filaments of similar dimension to those of platelet myosin. Chromaffin cell myosin binds to skeletal F-actin and forms characteristic arrowheads. Purified chromaffin myosin is poorly activated by actin similar to its smooth muscle or platelet counterpart suggesting that phosphorylation is required for its activation.[148] Uncharacterized factor(s) are present in chromaffin cells that are necessary for activation of chromaffin myosin by actin. One of these factors may be myosin light chain kinase; its gel filtration properties are consistent with the molecular weight of myosin light chain kinase. Calmodulin has been purified from the adrenal medulla.[149] Studies with antibodies to chromaffin myosin suggest that myosin is in the cytoplasm and may in part be localized near the plasma membrane.[148,150,151]

Phillips and Slater[152] demonstrated the presence of actin in the adrenal medulla. The actin was similar in peptide fingerprint and molecular weight to skeletal muscle actin. The actin accounted for 3 to 4% of extranuclear protein. The presence of actin specifically in the chromaffin cell was established later. Chromaffin actin was a mixture of β and γ actins with slightly greater amounts of the β form.[153] Immunofluorescence studies indicate that actin is most heavily concentrated near the plasma membrane.[154] Chromaffin cells also contain α-actinin that has been extracted from granules and identified by molecular weight, immunodiffusion, and immunoblotting.[150,155] A spectrin-like protein with two subunits of 240 and 235 kdaltons has been found.[156] A 92-kdalton polypeptide has been identified as gelsolin by its ability to bind actin and its cross reactivity with antimacrophage gelsolin.[153] Tropomyosin has been purified by a method used for the platelet protein and has been shown to be a tropomyosin by four different criteria.[157]

3. Interaction of Contractile Proteins with Chromaffin Granules

An indication that actin could associate with granules was obtained in immunofluorescence studies of cultured chromaffin cells. After 4 to 7 days in culture, cells exhibited a fine granular fluorescence that was similar to the pattern seen when the cells were stained with antidopamine β-hydroxalase, a granule constituent.[154] A similar punctate appearance of the cytoplasm is obtained when the immunofluorescent technique is applied with anti-α-actinin antibodies.[144,150] At the electron microscopic level, actin was localized by the protein A-gold technique and found to be near the granules' dense cores and in the cell web under the plasma membrane.[158] Three-dimensional stereo techniques indicate that a lattice of filaments connects the plasma membrane and the granule surface.

Isolated chromaffin granules have actin, spectrin, and α-actinin bound to their surface.[155,156,159-162] The actin associated with isolated granules can be extracted at low ionic strength, a condition which also removes α-actinin.[150,159,162] Cytochalasin B will decrease the amount of actin associated with granules.[160] The filaments associated with granules become shorter upon exposure to cytochalasin B. In the electron microscope, isolated granules have shown to be bound to both filament ends and to the center portion of microfilaments.[159,160] The filaments are of mixed polarity, i.e., some filaments have the barbed end associated with the membrane while others have the pointed end near the membrane. This contrasts with other systems in which only the barbed end is associated with the membrane.

Chromaffin granules contain nucleation sites for actin polymerization. α-Actinin has been purified from chromaffin granule membranes using similar extraction buffers to those used to remove α-actinin from the skeletal muscle Z-line.[150,155] This experiment indicates that α-actinin is not an integral membrane protein consistent with other systems. Antibodies to α-actinin reduce the number of actin nucleation sites indicating the presence of an actin-α-actinin complex on the membrane.[155] Spectrin, as mentioned, is also associated with the chromaffin cell. Treatment of chromaffin granule membranes with a nonionic detergent, Kyro EOB, releases the spectrin-like component of the membrane and reduces the actin-binding capacity of membranes by 50%.[156] These data indicate that there may two different types of actin nucleation sites on the chromaffin granule with possibly two different functions. It has been suggested that the spectrin-associated sites may be more sensitive to changes in cytoplasmic Ca^{2+}.[156] A 70-kdalton form of caldesmon has been recently identified in chromaffin cells.[85] This polypeptide had been shown to bind chromaffin granules with high affinity. Cytoplasmic gels assembled at high Ca^{2+} were depleted of both actin and the 70-kdalton polypeptide. The location of the 70-kdalton polypeptide was demonstrated by immunoperoxidase staining. Staining was most intense near the plasma membrane.

4. Physiological Studies of the Role of Contractile Proteins in Secretion

Two types of studies have been performed on intact chromaffin cells. Addition of cytochalasin B to these cells only weakly inhibited agonist-stimulated secretion.[163] Microinjection of antibodies to calmodulin had no effect on the ability of chromaffin cells to accumulate exogenous catecholamines but inhibited K^+-induced secretion.[164]

C. Secretion from Leukocytes

The word leukocytes encompasses a wide variety of different cell types but in studies of the contractile proteins of these cells it is easier to deal with them as a class with the caution that statements about one type of leukocyte may not necessarily be true about others. In addition, leukocytes display a wide variety of responses besides secretion that may be considered contractile including locomotion and phagocytosis. For a more complete review of contractile proteins in leukocyte function, see Southwick and Stossel.[165]

1. Contractile Proteins

A high percentage of the cytosolic proteins of leukocytes are contractile proteins. This system has been extensively studied by Southwick and Stossel.[165] Actin which is similar to other nonmuscle actins, represents 5 to 10% of total leukocyte protein.[166-168] Actin-binding protein (ABP) was first studied in macrophages where it comprises about 1% of total protein.[169] ABP is found in the areas of the cell that contain actin.[170,171] Profilin has been found in leukocytes and helps to account for the fact that 50% of leukocyte actin is monomeric.[172]

Acumentin has been found primarily in leukocytes where it may account for up to 7% of the protein of phagocytic cells.[172,173] Gelsolin accounts for about 1% of leukocyte protein.[174] Calmodulin has been isolated from leukocytes.[175] Myosin accounts for less than 1% of leukocyte protein.[166-168,176,177] Macrophage myosin 20,000 light chain is phosphorylated by endogenous kinase.[21] Phosphorylation of macrophage myosin was shown to be prerequisite for activation of the actin-activated ATPase of the myosin.

2. Localization of Leukocyte Contractile Proteins

When a leukocyte is attached to a surface, a veil of transparent material extends from the cell.[165] This veil is often associated with the side of the cell that points in the direction of movement. This region of the cell, called the hyaline cortex, excludes granules and other organelles and is observed to change in consistency during locomotion, degranulation, and phagocytosis. The contractile proteins of most leukocytes seem to concentrate in this region of the cell which is consistent with the view that in this region the force required for movement is developed. Very little information is available on the association of contractile proteins with granules.

3. Contractile Proteins and Leukocyte Secretion

Intact cell studies of contractile function in leukocytes have primarily relied on the introduction of cytochalasins into the cell. Cytochalasins, as would be expected, inhibit leukocyte movement and phagocytosis.[178] However, cytochalasins enhance agonist-induced secretion.[179] In fact, cytochalasin B is routinely used in studies of leukocyte secretion in order to optimize secretory products. However, since cytochalasin inhibits phagocytosis some of the effects of this agent may be caused by the lack of a phagocytic vacuole which is formed in control cells in response to some agonists. The contents of granules in cytochalasin-treated cells would be emptied into the medium rather than into the phagosome.

One study has examined the phosphorylation state of myosin after stimulation of lymphocytes. ^{32}P-labeled lymphocytes stimulated with an antibody to cell-surface immunoglobulin show increased incorporation of radioactivity into the 20-kdalton myosin light chain.[180]

4. Microtubules

When neutrophils were stimulated with C5a or immune complexes, the number of microtubules in each cell was found to increase.[181] Colchicine was shown to decrease microtubules in a dose-dependent manner. Colchicine had a similar effect on granule secretion but not as marked as its effect on microtubule depolymerization. When microtubules were fully disassembled secretion was only inhibited by 40%.

D. Platelet Secretion

1. Contractile Proteins

The platelet contains an abundance of contractile proteins. Actin accounts for 20 to 30% of total cellular protein.[36] Myosin accounts for 5% of cell protein.[14] Platelets also contain myosin light chain kinase,[15,16] tropomyosin,[51] α-actinin,[90] ABP,[74,75] gelsolin,[37] calmodulin, caldesmon,[84] vinculin,[98] talin,[102] and profilin.[55]

2. Localization of Platelet Contractile Proteins

Platelets are very small cells and are not suitable for study by immunofluorescent methods. Actin filaments are difficult to distinguish in electron micrographs of resting platelets. However, a layer of granule material is found next to the membrane beneath the microtubule ring.[183] Improved fixation techniques and the use of Triton X-100 to remove other features show that the submembrane granular material represents a ring of bundled actin filaments.[182,183] The central region of the cell contains a finer network of actin filaments. Dramatic changes in the organization of microfilaments occur when platelets are stimulated with agonists. The total amount of filaments in the cell appears to increase dramatically and filopodia are formed which contain bundles of actin filaments.[182,183] A dense ring of microfilaments is seen in the center of the cell surrounding the granules. This morphological change is again more apparent when Triton extracts of the cells are studied by electron microscopy. The location of other contractile proteins has not been studied in detail probably due to the small size of the cell.

3. Association of Contractile Proteins with Membranes and Granules

As mentioned activated platelets form filopodia-containing bundles of actin filaments. At the tip of these filopodia, actin appears to be close to the membrane. Nachmias and Asch[184] have shown that all the actin filaments of platelet filopodia have the same orientation with the barbed end of the filament being closest to the membrane. This orientation has been observed in many other similar systems and suggests a specific interaction between F-actin and membrane-associated elements. Actin is found in membrane preparations but the specificity of this interaction is questioned because actin tends to be a "sticky" protein and because microfilaments could easily become trapped with membrane. However, in two studies in which this issue was addressed, the conclusion was that the binding of actin was tight and specific.[185,186] The presence of spectrin-like and ankyrin-like proteins in platelets also speaks to a specific actin-membrane association. Little is known about the association of actin with platelets granules. An early report by Jockush et al.[187] indicated that α-actinin was present on platelet-dense granules but not α-granules.

4. Actin and Polymerization during Platelet Activation

The F-actin content of resting platelets has been determined to be 40 to 50% of total actin by the DNAase inhibition assay.[182] This is increased to 60 to 80% by platelet stimulation. This increase can be induced by a Ca^{2+} ionophore and is inhibited by elevation of cAMP. The change is rapid, i.e., is complete in 20 sec. Cytochalasins inhibit the agonist-induced increase in F-actin but do not appear to lower the resting level of F-actin. Carroll and co-workers have found that agonist-induced F-actin formation can be separated into two differently regulated processes.[188] Phorbol esters which activate protein kinase C cause formation of platelet filopodia without formation of a central contractile ring. Cytochalasins blocked formation of filopodia but did not inhibit thrombin-induced granule centralization. The effect of the addition of these agents on the amount of F-actin present in the cell was determined by the DNAase assay. The amount of F-actin formed by thrombin treatment was reduced but not returned to control levels suggesting that only part of the actin assembly was sensitive to cytochalasin.

5. Myosin Phosphorylation and Platelet Secretion

The relatively large amount of myosin in platelets as compared to other nonmuscle cells has allowed for a rather complete study of myosin phosphorylation during platelet activation. Stimulation of platelets with an agonist causes changes in two major phosphoproteins, a 40,000- or 47,000-dalton protein whose phosphorylation is catalyzed by protein kinase C and a 20,000-dalton protein shown to be a myosin light chain.[189-192] The level of phosphorylation can be conveniently quantitated by electrophoresis of total platelet protein on alkaline-urea PAGE.[193] Early studies suggested that there was a correlation between myosin phosphorylation and platelet secretion. However, we have shown that 50% of myosin is phosphorylated during the platelet shape change response when no secretion has occurred.[194] However, myosin phosphorylation associated with platelet shape change is transient and cells stimulated with collagen showed a second phase of phosphorylation that occurred during the time when secretion was also occurring. This experiment might indicate that myosin phosphorylation must be both quantitatively greater and of longer duration for secretion to occur. Nishizuka and co-workers suggested that secretion might require phosphorylation of both the 40,000 and 20,000 dalton proteins.[195,196] In their studies, stimulation of the cells with phorbol ester or synthetic diglycerides caused phosphorylation of the 40-kdalton polypeptide but not the 20-kdalton polypeptide. This agent caused secretion of serotonin but the secretion was slower and less extensive than that produced by physiologic agonists. Addition of the Ca^{2+} ionophore, A23187, caused phosphorylation of myosin light chain and greatly potentiated secretion so that the two agents together produce the sort of secretion seen with physiologic agonists. These data strongly suggest that the phosphorylation of both proteins was necessary for rapid secretion to occur. These data were also consistent with the observation that minimal phosphorylation of the 40-kdalton protein occurs when the cell's shape changed in the absence of secretion.[194] The fact that phosphorylation of myosin light chain and the 40-kdalton polypeptide are not sufficient for secretion is demonstrated by patients that show normal protein phosphorylation but abnormal secretion.[197]

6. Effects of Cytochalasins and Antimicrotubule Agents on Platelet Secretion

White studied the effect of various antimicrotubule drugs on platelet function.[198] These agents removed the microtubule ring and caused the discoid cell to become spherical. However, these drugs did not affect the ability of the cell to form filopodia, aggregate, or secrete granule constituents when the cells were challenged with an agonist.[199]

Cytochalasin B inhibits formation of platelet filopodia, a function that clearly requires polymerization of actin.[199] The effect of cytochalasin B on platelet secretion is biphasic.[200] At low concentrations in plasma, cytochalasin B (up to 10 μg/mℓ) potentiates the collagen-induced secretion of both dense and lysosomal constituents. Higher concentrations (20 to 40 μg/mℓ) inhibit collagen-induced secretion. The concentration dependence of the effect of cytochalasin on secretion was shifted to lower concentrations but was otherwise similar if the cells were washed free of plasma protein. In the studies of Carroll et al.[188] 10 μM cytochalasin B was found to inhibit filopodial development but not formation of a central microfilament ring. This concentration had little effect on serotonin secretion.

VII. MODELS OF THE ROLE OF CONTRACTILE PROTEINS IN SECRETION

The involvement of contractile proteins has been long assumed. This idea was based both on the presence of significant amounts of contractile proteins in secretory cells and the fact that similar to muscle contraction, secretion is triggered by an increase in intracellular Ca^{2+}. An early attempt to integrate contraction of actomyosin into a model for secretion was made by Durham.[201] Durham suggested that actin filaments were anchored at both granule and plasma membranes. Myosin filaments pulled the two sets of actin filaments and thus pulled the granule to the plasma membrane. Once the granule has been moved next to the plasma membrane, the contractile system exerts force onto the granule to aid in fusion of the two membranes. This mechanism is very much like the sliding filament mechanism of skeletal muscle with the two membranes acting as the z-line. In order to operate the barbed end of the actin filament it would have to be attached to both membranes. This has been established to be true in many cases for the association between plasma membrane and actin but the data on the association between granules and actin filaments are sparse. Of the systems discussed above, the best demonstration of a specific interaction between F-actin and secretory granules has been made in chromaffin cells. However, in that case the polarity of the attached filaments has not been clearly established. A basic problem with this model is that the same filaments that help to move the granule to the membrane would become a physical barrier as the granule approached the plasma membrane. Durham suggests that contraction and relaxation may alternate during secretory events, but does not indicate how this might occur.

Newer theories have incorporated new biochemical data about the properties of actin and associated proteins and new studies concerning the morphologic localization of contractile proteins. It has been suggested that models like that put forth by Durham, referred to as the "tow-rope" mechanism, are not realistic since simple diffusion is adequate to allow rapid movement of granules to the plasma membrane.[202] In addition, the design of secretory granules is such that in the case where secretion must be fast, granules are small allowing more rapid diffusion. Southwick and Stossel[165] have proposed that the dense cortical web of microfilaments provides a physical barrier to the interaction of the granule with the plasma membrane. This barrier must be removed for secretion to occur. Observations in the light microscope of neutrophils that are secreting suggest that the cytoplama of this region of the cell has been less viscous. It is suggested that a local rise in Ca^{2+} would occur in the region of the cell where secretion was occurring. The Ca^{2+} would cause disruption of the cortical web by weakening the interaction of actin and α-actinin and through the action of gelsolin. Gelsolin, upon binding Ca^{2+}, would disrupt the preformed microfilaments through its action as a severing protein. These actions of gelsolin would also disassemble any filaments associated with granules. In addition, myosin could interact with the filaments to remove them to other regions of the cell. It is suggested that in the case where the stimulus is evenly distributed over the plasma membrane that myosin would condense the F-actin filaments into foci that would leave large sections of the membrane exposed.

A similar hypothesis has been proposed by Burgoyne for chromaffin secretion.[203] Supporting this hypothesis is the fact that cytochalasin B does not usually inhibit but often potentiates secretion. Also, Boyles and Bainton who used a method for studying the attachment of filaments to the plasma membrane showed that the number of filaments associated with the membrane was less during secretion by neutrophils and that the filaments appeared to have aggregated into foci.[204] Problems for this theory arise from the observation that stimulated cells including neutrophils[205] in which the concentration of cytoplasmic Ca^{2+} is known to increase show increases in F-actin content when assayed by DNAase inhibition. An extension of this theory is that when a cell such as a neutrophil is stimulated, regions where actin filaments disassemble should contain high Ca^{2+} and regions where filaments are forming should be relatively low in Ca^{2+}. Thus, during phagocytosis, the pseudopodal regions of the cell which extend to surround the invading particle are regions where actin is polymerizing and should be regions of low Ca^{2+}. This question was recently investigated by Sawyer et al.[206] who used the fluorescent Ca^{2+} indicator quin 2 to measure Ca^{2+} concentrations in different parts of the neutrophil during phagocytosis of opsonized zymosan. They found that the intracellular Ca^{2+} concentration was highest in both the periphagosomal cytoplasm and in the pseudopodia. These data do not fit with the theory of Stossel.

Platelet secretion is different than secretion in other cells in that the granules are first drawn to the center of the cell. These granules appear to have been moved to the center by a contracting ring of microfilaments. A sliding filament interaction between the microfilaments of the ring and as yet unseen myosin filaments could drive a circumferential contraction and tightening of the ring forcing the granules to the center. Supporting this view are the experiments that show that concentrations of cytochalasin B which do not inhibit secretion, do not dismantle the contractile ring. Exocytosis occurs by fusion of the granule membrane with invaginations of the platelet membrane that have been called the "surface connecting system" (SCS). It is not clear what force causes this membrane system to enter the central granule region, but it is possible that the contractile proteins facilitate this movement. Movement of the granules to the center of the cell may allow the granule nearest to the SCS to fuse with the SCS. Subsequent fusion events with the SCS would occur at sites in the membrane that were originally part of a granule.

VIII. CONCLUDING REMARKS

The contractile system of cells is obviously tightly regulated since such a large number of proteins has evolved to control both the polymerization state of actin and the interaction between actin and myosin. However, more information is needed about specific location of actin, myosin, and regulatory proteins during the different physiologic events that occur during cell stimulation. Only in this way can we understand how an increase in cytoplasmic Ca^{2+} can trigger an apparent decrease in the number of microfilaments in one part of the cell while filaments appear to form in another region. It is likely that different microfilaments have different proteins associated with them which change their response to cell stimulation. For example, in platelets only the filaments that form filopodia are affected by addition of cytochalasin B. It is possible that one type of filament has a protein such as tropomyosin associated with it which alters its sensitivity to both cytochalasins and to agonist-dependent stimulation. In turn, this difference in sensitivity underlies differences in the reactivity of microfilaments to proteins such as gelsolin. While a good deal of information has been gathered in the last few years, it is evident that the function of the contractile system in secretion as well as other intracellular events will require several years more of intense investigation.

REFERENCES

1. **Alberts, B., Bray, D., Lewis, J., Raff, M., Roberts, K., and Watson, J. D.,** *Molecular Biology of the Cell,* Garland Press, New York, 1983.
2. **Korn, E. D.,** Actin polymerization and its regulation by proteins from nonmuscle cells, *Physiol. Rev.,* 62, 672, 1982.
3. **Lowey, S., Slayter, H. S., Weeds, A. G., and Baker, H.,** The substructure of the myosin molecule. I. Subfragments of myosin by enzymic degradation, *J. Mol. Biol.,* 42, 1, 1969.
4. **Sarker, S., Sreter, F. A., and Gergely, J.,** Light chains of myosins from white, red, and cardiac muscles, *Proc. Natl. Acad. Sci. U.S.A.,* 68, 946, 1971.
5. **Lowey, S. and Risby, D.,** Light chains from fast and slow muscle myosin, *Nature (London),* 234, 81, 1971.
6. **Weeds, A. G. and Pope, B.,** Chemical studies on light chains from cardiac and skeletal muscle myosin, *Nature (London),* 234, 85, 1971.
7. **Leger, J. and Focant, B.,** Low molecular weight components of cow smooth muscle myosins, *Biochim. Biophys. Acta,* 328, 166, 1973.
8. **Adelstein, R. S. and Conti, M. A.,** The characterization of contractile proteins from platelets and fibroblasts, *Cold Spring Harbor Symp. Quant. Biol.,* 27, 599, 1972.

9. **Trifaro, J.-M. and Ulpian, C.,** Isolation and characterization of myosin from the adrenal medulla, *Neuroscience,* 1, 483, 1976.

10. **Weeds, A. E.,** Light chains of myosin, *Nature (London),* 223, 1362, 1969.

11. **Huxley, H. E.,** Electron microscope studies on the structure of natural and synthetic protein filaments from striated muscle, *J. Mol. Biol.,* 7, 281, 1963.

12. **Niederman, R. and Pollard, T.,** Human platelet myosin. II. In vitro assembly and structure of myosin filaments, *J. Cell Biol.,* 67, 72, 1975.

13. **Clarke, M. and Spudich, J. A.,** Nonmuscle contractile proteins: the role of actin and myosin in cell motility and shape determination, *Annu. Rev. Biochem.,* 46, 797, 1977.

14. **Lucas, R. C., Rosenberg, S., Shafiq, S., Stracher, A., and Lawrence, J.,** The isolation and characterization of a cytoskeleton and a contractile apparatus from human platelets, in *Protides of Biological Fluids,* Peeters, H., Ed., Pergamon Press, New York, 1975, 465.

15. **Daniel, J. L. and Adelstein, R. S.,** Isolation and properties of platelet myosin light chain kinase, *Biochemistry,* 15, 2370, 1976.

16. **Dabroska, R. and Hartshorne, D. J.,** A Ca^{2+}- and modulator-activated myosin light chain kinase from non-muscle cells, *Biochem. Biophys. Res. Commun.,* 85, 1359, 1979.

17. **Hathaway, D. R. and Adelstein, R. S.,** Human platelet myosin light chain kinase requires the calcium binding protein calmodulin for activity, *Proc. Natl. Acad. Sci. U.S.A.,* 76, 1653, 1979.

18. **Walsh, M. P.,** Calmodulin-dependent myosin light chain kinases, *Cell Calcium,* 2, 333, 1981.

19. **Yerna, M. J., Dabrowska, R., Hartshorne, D. J., and Goldman, R. D.,** Calcium-sensitive regulation of actin-myosin interactions in baby hamster kidney (BHK-21) cells, *Proc. Natl. Acad. Sci. U.S.A.,* 76, 184, 1979.

20. **Scordillis, S. P. and Adelstein, R. S.,** A comparative study of the myosin light chain kinases from myoblast and muscle sources, *J. Biol. Chem.,* 253, 9041, 1978.

21. **Trotter, J. A. and Adelstein, R. S.,** Macrophage myosin. Regulation of actin-activated ATPase activity by phosphorylation of the 20,000 dalton light chain, *J. Biol. Chem.,* 254, 8781, 1979.

22. **Adelstein, R. S., Chacko, S., Barylko, B., Scordilis, S., and Conti, M. A.,** The role of myosin phosphorylation in the regulation of platelet and smooth muscle contractile proteins in, *Current Topics in Intracellular Reguation II. Contractile Systems in Non-muscle Tissue,* Perry, S. V., Margeth, A., and Adelstein, R. S., Eds., Elsevier/North Holland, Amsterdam 1976, 153.

23. **Pato, M. D. and Adelstein, R. S.,** Dephosphorylation of the 20,000 dalton light chain of myosin by two different phosphatases from smooth muscle, *J. Biol. Chem.,* 255, 6535, 1980.

24. **Pato, M. D. and Adelstein, R. S.,** Purification and characterization of a multisubunit phosphatase from turkey gizzard smooth muscle. The effect of calmodulin binding to myosin light chain kinase on dephosphorylation, *J. Biol. Chem.,* 258, 7047, 1983.

25. **Wolff, D. J. and Brostrom, C. O.,** Properties and functions of the calcium-dependent regulator protein, *Adv. Cyclic Nuceotide Res.,* 11, 27, 1979.

26. **Blumenthal, D. K. and Stull, J. T.,** Activation of skeletal muscle myosin light chain kinase by calcium (2 +) and calmodulin, *Biochemistry,* 19, 5608, 1980.

27. **Hartshorne, D. J., Siemankowski, R. F., and Aksoy, M. O.,** Ca regulation in smooth muscle and phosphorylation: some properties of the light chain kinase in *Muscle Contraction: Its Regulatory Mechanisms,* Ebashi, S., Maruyama, K., and Endo, M., Eds., Japan Scientist Societies Press, Tokyo; Springer-Verlag, Berlin, 1980, 287.

28. **Adelstein, R. S. and Klee, C. B.,** Smooth muscle myosin light chain kinase, in *Calcium and Cell Function,* Vol. 1, Cheung, Y., Ed., Academic Press, New York, 1980, 167.

29. **Adelstein, R. S., Conti, M. A., Hathaway, D. R., and Klee, C. B.,** Phosphorylation of smooth muscle light chain kinase by the catalytic subunit of adenosine 3':5'-monophosphate-dependent protein kinase, *J. Biol. Chem.,* 253, 8347, 1978.

30. **Hathaway, D. R., Eaton, C. R., and Adelstein, R. S.,** Phosphorylation of human platelet myosin light chain kinase by the catalytic subunit of cyclic AMP-dependent protein kinase, *Nature (London),* 291, 252, 1981.

31. **Conti, M. A. and Adelstein, R. S.,** The relationship between calmodulin binding and phosphorylation of smooth muscle myosin kinase by the catalytic subunit of 3':5' cAMP-dependent protein kinase, *J. Biol. Chem.,* 256, 3178, 1981.

32. **Miller, J. R., Silver, P. J., and Stull, J. T.,** The role of myosin light chain kinase phosphorylation in beta-adrenergic relaxation of tracheal smooth muscle, *Mol. Pharmacol.,* 24, 235, 1983.

33. **Nishizuka, Y.,** Calcium phospholipid turnover, and transmembrane signalling, *Phil. Trans. R. Soc. London Ser. B,* 302, 101, 1983.

34. **Naka, M., Nishikawa, M., Adelstein, R. S., and Hidaka, H.,** Phorbol ester-induced activation of human platelets is associated with protein kinase C phosphorylation of myosin light chains, *Nature (London),* 306, 490, 1983.

35. **Nishikawa, M., Hidaka, H., and Adelstein, R. S.,** Phosphorylation of smooth muscle heavy meromyosin by calcium-activated, phospholipid-dependent protein kinase. The effect on actin-activated MgATPase activity, *J. Biol. Chem.,* 258, 14069, 1983.

36. **Harris, H. E. and Weeds, A.,** Platelet actin: subcellular distribution and association with profilin, *FEBS Lett.,* 90, 84, 1978.

37. **Stossel, T. P., Chaponnier, C., Ezzel, R. M., Hartwig, J. H., Janmey, P. A., Kwiatkowski, D. J., Lind, S. E., Smith, D. B., Southwick, F. S., Yin, H. L., and Zaner, K. S.,** Nonmuscle actin binding proteins, *Annu. Rev. Cell Biol.,* 1, 353, 1985.

38. **Lazarides, E. and Weber, K.,** Actin antibody: the visualization of actin filaments in non-muscle cells, *Proc. Natl. Acad. Sci. U.S.A.,* 71, 2268, 1974.

39. **Vandekerckhove, J. and Weber, K.,** Chordate muscle actins differ distinctly from invertebrate muscle actins, *J. Mol. Biol.,* 179, 391, 1984.

40. **Garrels, J. L. and Gibson, W.,** Identification and characterization of multiple forms of actin, *Cell,* 9, 793, 1976.

41. **Vanderkerchhove, J. and Weber, K.,** At least six different actins are expressed in higher mammal: an analysis based on the amino acid sequence of the amino terminal peptide, *J. Mol. Biol.,* 126, 783, 1978.

42. **Korn, E. D.,** Biochemistry of actomyosin-dependent cell motility (a review), *Proc. Natl. Acad. Sci. U.S.A.,* 75, 588, 1978.

43. **Harris, H.,** Regulation of motile activity in platelets, in *Platelets in Biology and Pathology,* Vol. 2, Gordon, J. L., Ed., Elsevier/North Holland, Amsterdam, 1981, 473.

44. **Kabsch, W., Mannherz, H. G., and Suck, D.,** Three-dimensional structure of the complex of actin and DNase I at 4.5 A resolution, *EMBO J.,* 4, 2113, 1985.

45. **Oosawa, F. and Kasai, M.,** Actin, in *Subunits in Biological Systems,* Timasheff, S. N. and Fasman, G. D., Eds., New York, Marcel Dekker, 1971, 261.

46. **Daniel, J. L., Robkin, L., Molish, I. R., and Holmsen, H.,** Determination of the ADP concentration available to participate in energy metabolism in an actin-rich cell, the platelet, *J. Biol. Chem.,* 254, 7878, 1979.

47. **Huxley, H. E.,** Electron microscope studies on the structure of natural and synthetic proteins filaments from striated muscle, *J. Mol. Biol.,* 7, 281, 1963.

48. **Pollard, T. D. and Cooper, J. A.,** Actin and actin-binding proteins. A critical evaluation of mechanisms and functions, *Annu. Rev. Biochem.,* 55, 987, 1986.

49. **Pollard, T. D. and Mooseker, M.,** Direct measurement of actin polymerization rate constants by electron microscopy of actin filaments nucleated by isolated microvillus cores, *J. Cell Biol.,* 88, 654, 1981.

50. **Gordon, D. J., Boyer, J. L., and Korn, E. D.,** Comparative biochemistry of non-muscle actins, *J. Biol. Chem.,* 252, 8300, 1977.

51. **Smillie, L. B.,** Structure and function of tropomyosins from muscle and non-muscle, *Trends Biochem. Sci.,* 4, 151, 1981.

52. **Cote, G. P. and Smillie, L. B.,** The interaction of equine platelet tropomyosin with skeletal muscle actin, *J. Biol. Chem.,* 256, 7257, 1981.

53. **Lazarides, E.,** Two general classes of cytoplasmic actin filaments in tissue culture cells: the role of tropomyosin, *Supramol. Struct.,* 5, 531, 1976.

54. **Carlson, L., Nystrom, L.-E., Sundquist, I., Markey, F., and Lindberg, U.,** Actin polymerizability is influenced by profilin, a low molecular weight protein in non-muscle cells, *J. Mol. Biol.,* 115, 465, 1977.

55. **Markey, F., Lindberg, U., and Eriksson, L.,** Human platelets contain profilin, a potential regulator of actin polymerizability, *FEBS Lett.,* 88, 75, 1978.

56. **Blikstad, I., Sundquist, I., and Eriksson, S.,** Isolation of profilin and profilactin from calf thymus and brain, *Eur. J. Biochem.,* 105, 425, 1980.

57. **Fattoum, A., Roustan, C., Feinberg, J., and Prandel, L. A.,** Biochemical evidence for low molecular weight protein (profilin-like protein) in hog thyroid gland and its involvements in actin polymerization, *FEBS Lett.,* 118, 237, 1980.

58. **DiNubile, M. J. and Southwick, F. S.,** Effects of macrophage profilin on actin in the presence of acumentin and gelsolin, *J. Biol. Chem.,* 260, 7402, 1985.

59. **Mockrin, S. C. and Korn, E. D.,** Acathamoeba profilin interacts with G-actin to increase the rate of exchange of actin-bound adenosine-5'-triphosphate, *Biochemistry,* 19, 5359, 1980.

60. **Lassing, I. and Lindberg, U.,** Specific interaction between phosphatidylinositol 4,5-biphosphate and profilactin, *Nature (London),* 314, 472, 1985.

61. **Mannherz, H. G., Goody, R. S., Konrad, M., and Novak, E.,** The interaction of bovine pancreatic deoxyribonuclease I and skeletal muscle actin, *Eur. J. Biochem.,* 104, 367, 1980.

62. **Pinder, J. C. and Gratzer, W. B.,** Investigation of the actin-deoxyribonuclease interaction using a pyrene-conjugated actin derivative, *Biochemistry,* 21, 4886, 1982.

63. **Mannherz, H. G., Leigh, J. B., Leberman, R., and Prang, H.,** A specific 1:1 G-actin:DNAase I complex formed by the action of DNAase I on F-actin, *FEBS Lett.,* 60, 34, 1980.

64. **Hitchcock, S. E., Carlsson, L., and Lindberg, U.,** Depolymerization of F-actin by deoxyribonuclease I, *Cell*, 7, 531, 1976.

65. **Blikstad, I., Markey, F., Carlsson, L., Persson, T., and Lindberg, U.,** Selective assay of monomeric and filamentous actin in cell extracts, using inhibition of deoxyribonuclease I, *Cell*, 15, 935, 1978.

65a. **Yin, H. L. and Stossel, T. P.,** Control of cytoplasmic actin gel-sol transformation by gelsolin, a calcium-dependent regulatory protein, *Nature (London)*, 281, 583, 1979.

66. **Bryan, J. and Kurth, M.,** Actin-gelsolin interactions. Evidence for two actin-binding sites, *J. Biol. Chem.*, 259, 7480, 1984.

67. **Culuccio, L. M., Kurth, M. C., and Bryan, J.,** Gelsolin, a multifunctional actin associated protein, *J. Cell Biol.*, 99, 307a, 1984.

68. **Jamney, P. A., Chaponnier, C., Lind, S. E., Zaner, K. S., Stossel, T. P., and Yin, H. L.,** Interactions of gelsolin and gelsolin actin complexes with actin. Effects of Ca^{2+} on actin nucleation, filament severing and end blocking, *Biochemistry*, 24, 3714, 1985.

69. **Yin, H. L., Kwiatkowski, D. J., Mile, J. E., and Cole, F. S.,** Structure and biosynthesis of cytoplasmic and secreted variants of gelsolin, *J. Biol. Chem.*, 259, 5271, 1984.

70. **Southwick, F. S., Tatsumi, N., and Stossel, T. P.,** Acumentin an actin-modulating protein of rabbit pulmonary macrophages, *Biochemistry*, 21, 6321, 1982.

71. **Southwick, F. S. and Stossel, T. P.,** Isolation of an inhibitor of actin polymerization from human polymorphonuclear leukocytes, *J. Biol. Chem.*, 256, 3030, 1981.

72. **Maruyama, K. and Sakai, H.,** Cell B-actinin, an accelerator of actin polymerization, isolated from rat kidney cytosol, *J. Biochem.*, 89, 1337, 1981.

73. **Hartwig, J. H. and Stossel, T. P.,** Isolation and properties of actin, myosin, and a new actin-binding protein in rabbit aveolar macrophages, *J. Biol. Chem.*, 250, 5696, 1975.

74. **Lucas, R. C., Gallagher, M., and Stracher, A.,** Actin and actin-binding protein in platelets, in *Contractile Systems in Non-muscle Tissue*, Perry, S. V. et al., Eds., Elsevier/North Holland, Amsterdam, 1979, 133.

75. **Schollmeyer, J. V., Rao, G. H. R., and White, J. G.,** An actin-binding protein in human platelets, *Am. J. Pathol.*, 93, 433, 1978.

76. **Schloss, J. A. and Goldman, R. D.,** Isolation of a high molecular weight actin-binding protein from baby hamster kidney (BKH-21) cells, *Proc. Natl. Acad. Sci. U.S.A.*, 76, 4484, 1979.

77. **Hartwig, J. H. and Stossel, T. P.,** Structure of macrophage actin-binding protein molecules in solution and interaction with actin filaments, *J. Mol. Biol.*, 145, 563, 1981.

78. **Wallach, D., Davies, P. J. A., and Pastan, I.,** Purification of mammalian filamin. Similarity to high molecular weight actin-binding protein in macrophages, platelets, fibroblasts, and other tissues, *J. Biol. Chem.*, 253, 3328, 1978.

79. **Collier, N. C. and Wang, K.,** Human platelet P235: a high M_r protein which restricts the length of actin filaments, *FEBS Lett.*, 143, 205, 1982.

80. **Hartwig, J. H., Tyler, J., and Stossel, T. P.,** Actin-binding promotes bipolar and perpendicular branching of actin filaments, *J. Cell Biol.*, 87, 841, 1980.

81. **Qing-Qi, Z., Rosenberg, S., Lawerence, J., and Stracher, A.,** Role of actin binding protein phosphorylation in platelet cytoskeletal assembly, *Biochem. Biophys. Res. Commun.*, 118, 508, 1984.

82. **Bretscher, A.,** Smooth muscle caldesmon. Rapid purification and F-actin cross-linking properties, *J. Biol. Chem.*, 259, 12873, 1984.

83. **Sobue, K., Morimoto, K., Kanda, K., Maruyama, K., and Kakiuchi, S.,** Reconstitution of Ca^{2+}-sensitive gelation of actin filaments with filamin, caldesmon, and calmodulin, *FEBS Lett.*, 138, 289, 1982.

84. **Pho, D. B., Desbruyeres, E., DerTerrossian, E., and Olomucki, A.,** Cytoskeletons of ADP- and thrombin-stimulated blood platelets, *FEBS Lett.*, 202, 117, 1986.

85. **Burgoyne, R. D., Cheek, T. R., and Norman, K.-M.,** Identification of a secretory granule-binding protein as caldesmon, *Nature (London)*, 319, 68, 1986.

86. **Podlubnaya, Z. A., Tskhovrebova, L. A., Zaalishvili, M. M., and Stefanenko, G. A.,** Electron microscopic study of alpha-actinin, *J. Mol. Biol.*, 92, 357, 1975.

87. **Suzuki, A., Goll, D. E., Singh, I., Allen, R. E., Robson, R. M., and Stromer, M. E.,** Some properties of purified skeletal muscle alpha-actinin, *J. Biol. Chem.*, 251, 6870, 1976.

88. **Jockusch, B. M. and Isenberg, G.,** Interaction of alpha-actinin and vinculin with actin: opposite effects on filament network formation, *Proc. Natl. Acad. Sci. U.S.A.*, 78, 3005, 1981.

89. **Blikstad, I., Eriksson, S., and Carlsson, L.,** Alpha-actinin promotes polymerization of actin profilactin, *Eur. J. Biochem.*, 109, 317, 1980.

90. **Landon, F. and Olomucki, A.,** Isolation and physico-chemical properties of blood alpha-actinin, *Biochim. Biophys. Acta*, 742, 129, 1983.

91. **Rosenberg, S., Stracher, A., and Burridge, K.,** Isolation and characterization of a calcium-sensitive alpha-actinin-like protein from human platelet cytoskeletons, *J. Biol. Chem.*, 256, 12986, 1981.

92. **Geiger, B.,** A 130K protein from chicken gizzard: its localization at the termini of microfilament bundles in cultured chicken cells, *Cell*, 18, 193, 1979.

93. **Geiger, B., Tokuyasu, K. T., Dutton, A. H., and Singer, S. J.,** Vinculin an intracellular protein localized at specialized sites where microfilaments terminate at cell membranes, *Proc. Natl. Acad. Sci. U.S.A.,* 77, 4127, 1980.

94. **Burridge, K. and McCullough, L.,** The association of alpha-actinin with the plasma membrane, *J. Supramol. Struct.,* 13, 53, 1980.

95. **Jockusch, B. M. and Isenberg, G.,** Vinculin and alpha-actinin: interaction with actin and effect on microfilament network formation, *Cold Spring Harbor Symp. Quant. Biol.,* 46, 613, 1982.

96. **Burridge, K. and Feramisco, J. R.,** Alpha-actinin and vinculin from nonmuscle cells: calcium-sensitive interactions with actin, *Cold Spring Harbor Symp. Quant. Biol.,* 46, 587, 1982.

97. **Lin, S., Wilkins, J. A., Cribbs, D. H., Grumet, M., and Lin, D. C.,** Proteins and complexes that affect actin-filament assembly and interactions, *Cold Spring Harbor Symp. Quant. Biol.,* 46, 625, 1982.

98. **Koteliansky, V. E., Gneushev, G. N., Glukhova, M. A., Venyaminov, S. Y., and Muszbek, L.,** Identification and isolation of vinculin from platelets, *FEBS Lett.,* 165, 26, 1984.

99. **Sefton, B. M., Hunter, T., Nigg, E. A., Singer, S. J., and Walter, G.,** Cytoskeletal targets for viral transforming proteins with tyrosine protein kinase activity, *Cold Spring Harbor Symp. Quant. Biol.,* 46, 939, 1982.

100. **Burridge, K. and Connell, L.,** A new protein of adehesion plaques and ruffling of membranes, *J. Cell Biol.,* 97, 359, 1983.

101. **Burridge, K. and Mangeat, P.,** The interaction between vinculin and talin, *Nature (London),* 308, 744, 1984.

102. **O'Halloran, T., Beckerle, M. C., and Burridge, K.,** Identification of talin as a major cytoplasmic protein implicated in platelet activation, *Nature (London),* 317, 449, 1985.

103. **Berkerle, M. C., O'Halloran, T., and Burridge, K.,** Demonstration of a relationship between talin and P235, a major substrate of Ca^{2+}-dependent protease in platelets, *J. Cell. Biochem.,* 30, 259, 1986.

104. **Bennett, V.,** The molecular basis for membrane-cytoskeletal association in human erythrocytes, *J. Cell. Biochem. Cell Recognition,* 18, 513, 1982.

105. **Schnaus, E., Booth, S., Hallaway, B., and Rosenberg, A.,** The elasticity of spectrin-actin gels at high protein concentration, *J. Biol. Chem.,* 260, 3724, 1985.

106. **Davies, G. E. and Cohen, C. M.,** Platelets contain proteins immunologically related to red cell spectrin and band 4.1, *Blood,* 62 (Suppl. 1), 254a, 1983.

107. **Levine, J. and Willard, M.,** Fodrin: axonally transported polypeptides associated with the internal periphery of many cells, *J. Cell Biol.,* 90, 631, 1981.

108. **Glenney, J. R., Glenney, P., Osborn, M., and Weber, K.,** A high molecular weight F-actin and calmodulin binding protein with spectrin-related morphology is a major constituent of the terminal web microfilament organization of isolated intestinal brush borders, *Cell,* 28, 843, 1982.

109. **Bennett, V.,** Immunoreactive forms of human erythrocyte ankyrin are present in diverse cells and tissues, *Nature (London),* 281, 597, 1979.

110. **Adelstein, R. S. and Eisenberg, E.,** Regulation and kinetics of the actin-myosin-ATP interaction, *Annu. Rev. Biochem.,* 49, 921, 1980.

111. **Adelstein, R. S. and Conti, M. A.,** Phosphorylation of platelet myosin increases actin-activated myosin ATPase activity, *Nature (London),* 256, 597, 1975.

112. **Aksoy, M. O., Williams, D., Sharkey, E. M., and Hartshorne, D. J.,** A relationship between the Ca^{2+} sensitivity and phosphorylation of gizzard actomyosin, *Biochem. Biophys. Res. Commun.,* 78, 1263, 1976.

113. **Chacko, S., Conti, M. A., and Adelstein, R. S.,** Effect of phosphorylation of smooth muscle myosin on actin activation and Ca^{2+} regulation, *Proc. Natl. Acad. Sci. U.S.A.,* 74, 129, 1977.

114. **Sellers, J. R., Pato, M. D., and Adelstein, R. S.,** Reversible phosphorylation of smooth muscle myosin, heavy meromyosin, and platelet myosin, *J. Biol. Chem.,* 256, 13137, 1981.

115. **Lebowitz, E. A. and Cooke, R.,** Contractile properties of actomyosin from human blood platelets, *J. Biol. Chem.,* 253, 5443, 1978.

116. **Salganicoff, L., Loughnane, M., Sevy, R. W., and Russo, M.,** The platelet strip. I. A low fibrin contractile model of thrombin-activated platelets, *Am. J. Physiol.,* 248, C279, 1985.

117. **Salganicoff, L. and Sevy, R.,** The platelet strip. II. Pharmacomechanical coupling in thrombin-activated human platelets, *Am. J. Physiol.,* 249, C288, 1985.

118. **Daniel, J. L., Molish, I. R., Holmsen, H., and Salganicoff, L.,** Phosphorylation of myosin light chain in intact platelets: possible role in platelet secretion and clot retraction, *Cold Spring Harbor Conf. Cell Prolif.,* 8, 913, 1981.

119. **Bromberg, M. E., Sevy, R. W., Daniel, J. L., and Salganicoff, L.,** The role of myosin phosphorylation in the contractility of a platelet aggregate, *Am. J. Physiol.,* 18, C297, 1985.

120. **Suzuki, H., Onsihi, H., and Wantabe, S. K.,** Structure and function of gizzard myosin, *J. Biochem. (Tokyo),* 84, 1529, 1978.

121. **Scholey, J. M., Taylor, K. A., and Kendrick-Jones, J.,** Regulation of nonmuscle myosin assembly by calmodulin-dependent light chain kinase, *Nature (London),* 287, 233, 1980.

122. **Somylo, A. V., Butler, T. M., Bond, M., and Somylo, A. P.,** Myosin filaments have nonphosphorylated light chains in relaxed smooth muscle, *Nature (London),* 294, 567, 1981.

123. **Cande, W. Z., Tooth, P. J., and Kendrick-Jones, J.,** Regulation of contraction and thick filament assembly-disassembly in glycerinated vertebrate smooth muscle cells, *J. Cell Biol.,* 97, 1062, 1983.

124. **Dean, P. M.,** Ultrastructural morphometry of the pancreatic β-cell, *Diabetologia,* 9, 115, 1973.

125. **Lacy, P. E., Howell, S. L., Young, D. E., and Fink, C. J.,** New hypothesis of insulin secretion, *Nature (London),* 219, 1177, 1968.

126. **Howell, S. L., Hii, C. S. T., Shaikh, S., and Tyhurst, M.,** Effects of taxol and nocodazole on insulin secretion from isolated rat islets of Langerhans, *Biochem. J.,* 148, 237, 1982.

127. **Lacy, P. E., Walker, M. M., and Fink, C. J.,** Perfusion of isolated rat islets in vitro. Participation of the microtubular system in the phasic pattern of insulin release, *Diabetes,* 21, 987, 1972.

128. **Suprenant, K. A. and Dentler, W. L.,** Association between endocrine pancreatic secretory granules and in vitro assembled microtubules is dependent upon microtubule-associated proteins, *J. Cell Biol.,* 93, 164, 1982.

129. **Pipeleers, D. G., Harnie, N. Heylen, L., and Wauters, G.,** Microtubule interaction in islets of Langerhans, in *Biochemistry and Biophysics of the Pancreatic B Cell. European Workshop, Brussels,* Malaisse, W. J. and Taljedal, I. B., Eds., Georg Thieme Verlag, Stuttgart, 1979, 163.

130. **Gabbiani, G., Malaisse-Lagae, F., Blondel, B., and Orci, L.,** Actin in pancreatic islet cells, *Endocrinology,* 95, 1630, 1974.

131. **Van Obberghen, E., Sommers, G., Devis, G., Vaughn, G. D., Malaisse, F., Orci, L., and Malaisse, W. J.,** Dynamics of insulin release of microtubular-microfilamentous system. I. Effect of cytochalasin B, *J. Clin. Invest.,* 32, 1041, 1973.

132. **Howell, S. L. and Tyhurst, M.,** Regulation of actin polymerization in rat islets of Langerhans, *Biochem. J.,* 192, 381, 1980.

133. **Swanston-Flatt, S. K., Carlsson, L., and Gylie, E.,** Actin filament formation in pancreatic B cells during glucose stimulation of insulin secretion, *FEBS Lett.,* 117, 299, 1980.

134. **Ostlund, R. E., Lueng, J. T., and Kipnis, D. M.,** Myosins of secretory tissues, *J. Cell Biol.,* 77, 827, 1977.

135. **McDonald, M. J. and Kowluru, A.,** Calcium-calmodulin dependent myosin phosphorylation by pancreatic islets, *Diabetes,* 31, 566, 1982.

136. **Penn, E. J., Brocklehurst, K. W., Sopwith, A. M., Hales, C. N., and Hutton, J. C.,** Ca^{2+}-calmodulin dependent myosin light chain phosphorylation activity in insulin secreting tissues, *FEBS Lett.,* 139, 4, 1982.

137. **McDonald, M. J. and Kowluru, A.,** Calcium activated factors in pancreatic islets that inhibit actin polymerization, *Arch. Biochem. Biophys.,* 219, 489, 1982.

138. **Howell, S. L. and Tryhurst, M.,** Actomyosin interactions with insulin storage granules in vitro, *Biochem. J.,* 206, 157, 1982.

139. **Griffith, L. M. and Pollard, T. D.,** Evidence for actin filament-microtubule interaction mediated by microtubule-associated protein, *J. Cell Biol.,* 78, 958, 1978.

140. **Nishida, E., Kuwaki, T., and Sakai, H.,** Phosphorylation of microtubule-associated protein (MAP's) and pH of the medium control interaction between MAP's and actin filaments, *J. Biochem.,* 90, 975, 1981.

141. **Howell, S. L. and Tryhurst, M.,** Insulin secretion: the effector system, *Experientia,* 40, 1098, 1984.

142. **Lacy, P. E., Fink, E. H., and Codilla, R. C.,** Cinemicrographic studies on granule movement in monolayer culture of islet cells, *Lab. Invest.,* 33, 570, 1975.

143. **Somers, G., Blondel, B., Orci, L., and Malaisse, W. J.,** Motile events in pancreatic endocrine cells, *Endocrinology,* 104, 255, 1979.

144. **Bader, M.-F., Bernier-Valentin, F., Rousset, B., and Aunis, D.,** The adrenal paraneurone: tubulin organization, *Can. J. Physiol. Pharmacol.,* 62, 502, 1984.

145. **Bader, M.-F., Cielski-Tieska, J., Thierse, D., Hesketh, J. E., and Aunis, D.,** Immunocytochemical study of microtubules in chromaffin cells in culture and evidence that tubulin is not an integral protein of the chromaffin granule membrane, *J. Neurochem,* 37, 917, 1981.

146. **Bernier-Valentin, F., Aunis, D., and Rousset, B.,** Evidence for tubulin-binding sites on cellular membranes: plasma membranes, mitochondrial membranes and granule membranes, *J. Cell Biol.,* 97, 209, 1983.

147. **Trifaro, J.-M., Collier, B., Lastowecka, A., and Stern, D.,** Inhibition by colchicine and by vinblastine of acetylcholine-induced catecholine release from the adrenal glands: an anticholinergic action, not an effect upon microtubules, *Mol. Pharmacol.,* 8, 264, 1972.

148. **Cruetz, C. E.,** Isolation, characterization and localization of bovine adrenal medullary myosin, *Cell Tissue Res.,* 178, 17, 1977.

149. **Kuo, I. C. Y. and Coffee, C. J.,** Purification and characterization of a troponin-C like protein from bovine adrenal medulla, *J. Biol. Chem.,* 231, 1603, 1976.

150. **Aunis, D., Guerold, B., Bader, M.-F., and Cieselski-Treska, J.,** Immunocytochemical and biochemical demonstration of contractile proteins in chromaffin cells in culture, *Neuroscience,* 5, 2261, 1980.

151. **Bader, M.-F., Garcia, A. G., Cieselski-Treska, J., Thierse, D., and Aunis, D.,** Contractile proteins in chromaffin cells, *Prog. Brain Res.,* 58, 21, 1983.

152. **Phillips, J. and Slater, A.,** Actin in the adrenal medulla, *FEBS Lett.,* 56, 327, 1975.

153. **Lee, R. W. H., Mushynski, W. E., and Trifaro, J.-M.,** Two forms of cytoplasmic actin is in adrenal chromaffin cells, *Neuroscience,* 4, 843, 1979.

154. **Lee, R. W. H. and Trifaro, J.-M.,** Characterization of anti-actin antibodies and their use in immunocytochemical studies on the localization of actin in adrenal chromaffin cells, *Neuroscience,* 6, 2087, 1981.

155. **Bader, M.-F. and Aunis, D.,** The 97-kd alpha-actinin-like protein in chromaffin granule membranes from adrenal medulla: evidence for localization on the cytoplasmic surface and for binding to actin filaments, *Neuroscience,* 8, 165, 1983.

156. **Aunis, D. and Perrin, D.,** Chromaffin granule-F-actin interactions and spectrin-like protein of subcellular organelles: a possible relationship, *J. Neurochem.,* 42, 1558, 1984.

157. **Cote, A., Doucet, J.-P., and Trifaro, J.-P.,** Chromaffin cell tropomyosin, in *Molecular Biology of Peripheral Catecholamine Storing Tissues,* Colmar, France 1984, 100.

158. **Bendayan, M., Marceau, N., Beaudoin, A. R., and Trifaro, J.-M.,** Immunocytochemical localization of actin in the pancreatic exocrine cell, *J. Histochem. Cytochem.,* 30, 1075, 1982.

159. **Burridge, K. and Phillips, J. M.,** Association of actin and myosin with secretory granule membranes, *Nature (London),* 254, 526, 1975.

160. **Wilkins, J. A. and Lin, S.,** Association of actin with chromaffin granule membranes and the effect of cytochalasin B on the polarity of actin filament elongation, *Biochim. Biophys. Acta,* 642, 55, 1981.

161. **Fowler, V. M. and Pollard, H. B.,** Chromaffin granule membrane-F-actin interactions are calcium sensitive, *Nature (London),* 295, 336, 1982.

162. **Kao, L.-S. and Westhead, E. W.,** Binding of actin to chromaffin granules and mitochondrial fractions of adrenal medulla, *FEBS Lett.,* 173, 119, 1984.

163. **Schneider, A. S., Cline, H. T., Rosenheck, K., and Sonenberg, M.,** Stimulus-secretion coupling in isolated chromaffin cells: calcium channel activation and possible role of cytoskeletal elements, *J. Neurochem.,* 37, 567, 1981.

164. **Knight, D. E. and Baker, P. F.,** Calcium-dependence of catecholamine release from bovine adrenal medullary cells after exposure to intense electric fields, *J. Membrane Biol.,* 68, 107, 1982.

165. **Southwick, F. S. and Stossel, T. P.** Contractile proteins in leukocyte function, *Semin. Hematol.,* 20, 305, 1983.

166. **Boxer, L. A. and Stossel, T. P.,** Interactions of actin, myosin, and an actin-binding protein of chronic myelogenous leukemia leukocytes, *J. Clin. Invest.,* 57, 964, 1976.

167. **Fechheimer, M. and Cebra, J. J.,** Isolation and characterization of actin and myosin from B-lymphocyte guinea pig leukemia cells, *J. Immunol.,* 122, 2590, 1979.

168. **Hartig, J. H. and Stossel, T. P.,** Interactions of actin, myosin, and an actin-binding protein in rabbit pulmonary macrophages. III. Effects of cytochalasin B, *J. Cell Biol.,* 71, 295, 1976.

169. **Hartig, J. H. and Stossel, T. P.,** Isolation and properties of actin, myosin, and a new actin-binding protein in rabbit aveolar macrophages, *J. Biol. Chem.,* 250, 5699, 1975.

170. **Thorstensson, R., Utter, G., Norber, R., Fagraeus, A., Hartwig, J. H., Yin, H. L., and Stossel, T. P.,** Distribution of actin, myosin, actin-binding protein and gelsolin in cultured lymphoid cells, *Exp. Cell Res.,* 140, 395, 1982.

171. **Valerius, N. H., Stendahl, O. I., Hartwig, J. H., and Stossel, T. P.,** Distribution of actin-binding protein and myosin in polymorphonuclear leukocytes during locomotion and phagocytosis, *Cell,* 24, 195, 1981.

172. **Southwick, F. S. and Stossel, T. P.,** Isolation of an inhibitor of actin polymerization from human polymorphonuclear leukocytes, *J. Biol. Chem.,* 256, 3030, 1981.

173. **Southwick, F. S., Tatsumi, N., and Stossel, T. P.,** Acumentin, an actin-modulating protein of rabbit pulmonary macrophages, *Biochemistry,* 24, 6321, 1982.

174. **Yin, H. L. and Stossel, T. P.,** Control of cytoplasmic actin gel-sol transformation by gelsolin, a calcium-dependent regulary protein, *Nature (London),* 218, 586, 1979.

175. **Jamieson, G. A., Jr. and Vanaman, T. C.,** Isolation and characterization of calmodulin from a murine macrophage-like cell line, *J. Immunol.,* 125, 1171, 1980.

176. **Stossel, T. P.,** Phagocytosis, *N. Engl. J. Med.,* 290, pp. 717, 774, 833, 1974.

177. **Stossel, T. D. and Pollard, T. D.,** Myosin in polymorphonuclear leukocytes, *J. Biol. Chem.,* 248, 8288, 1973.

178. **Axline, S. G. and Reaven, E. P.,** Inhibition of phagocytosis and plasma membrane mobility of the cultivated macrophage by cytochalasin B. Role of subplasmalemmal microfilaments, *J. Cell Biol.,* 62, 647, 1974.

179. **Zurier, R. B., Hoffstein, S., and Weissman, G.,** Cytochalasin B: effect on lysosomal enzyme release from human leukocytes, *Proc. Natl. Acad. Sci. U.S.A.,* 70, 844, 1973.

180. **Fechheimer, M. and Cebra, J. J.,** Phosphorylation of lymphocyte myosin catalyzed in vitro and in intact cells, *J. Cell Biol.,* 93, 261, 1982.

181. **Hoffstein, S., Goldstein, I. M., and Weissman, G.,** Role of microtubule assembly in lysosomal enzyme secretion from human polymorphonuclear leukocytes, A reevaluation, *J. Cell Biol.,* 73, 242, 1977.

182. **Fox, J. E. B. and Phillips, D. R.,** Polymerization and organization of actin filaments within platelets, *Semin. Hematol.,* 20, 243, 1983.

183. **Tuszynski, G. P., Daniel, J. L., and Stewart, G.,** Association of proteins with the platelet cytoskeleton, *Semin. Hematol.,* 22, 303, 1985.

184. **Nachmias, V. T. and Asch, A.,** Regulation and polarity: results with myxomycete plasmodium and with human platelets, *Cold Spring Harbor Conf. Cell Prolif.,* 3, 771, 1976.

185. **Crawford, N.,** Platelet microfilaments and microtubules, in *Platelets in Biology and Pathology,* Gordon, J. L., Ed., Elsevier/North Holland, Amsterdam, 1976, 126.

186. **Davies, G. E.,** Association of actin with the platelet membrane, *Biochim. Biophys. Acta,* 772, 149, 1984.

187. **Jockush, B. M., Burger, M. M., DaPrada, M., Richards, J. G., Chaponnier, C., and Gabbiani, G.,** Alpha-actinin attached to the membranes of secretory vesicles, *Nature (London),* 270, 628, 1977.

188. **Carroll, R. C., Bulter, R. G., Morris, P. A., and Gerrard, J. M.,** Separable assembly of platelet pseudopodal and contractile cytoskeletons, *Cell,* 30, 385, 1982.

189. **Lyons, R. M., Stanford, N., and Majerus, P. W.,** Thrombin-induced protein phosphorylation in human platelets, *J. Clin. Invest.,* 56, 924, 1975.

190. **Haslam, R. J. and Lynham, J. A.,** Relationship between phosphorylation of blood platelet proteins and secretion of platelet granule constituents. I. Effect of different aggregating agents, *Biochem. Biophys. Res. Commun.,* 7, 714, 1977.

191. **Nishizuka, Y.,** Calcium, phospholipid turnover, and transmembrane signalling, *Phil. Trans. R. Soc. London Ser. B,* 302, 101, 1983.

192. **Daniel, J. L., Holmsen, H., and Adelstein, R. S.,** Thrombin-stimulated myosin phosphorylation in intact platelets and its possible involvement in secretion, *Thromb. Haemostosis,* 38, 984, 1977.

193. **Daniel, J. L., Molish, I. R., and Holmsen, H.,** Myosin phosphorylation in intact platelets, *J. Biol. Chem.,* 256, 7510, 1981.

194. **Daniel, J. L., Molish, I., Rigmaiden, M., and Stewart, G.,** Evidence for a role of myosin phosphorylation in the initiation of the platelet shape change response, *J. Biol. Chem.,* 259, 9826, 1984.

195. **Yamanshi, J., Takai, Y., Kaibuchi, K., Sano, K., Castagna, M., and Nishizuka, Y.,** Syngeristic functions of phorbol ester and calcium in serotonin release from human platelets, *Biochem. Biophys. Res. Commun.,* 112, 778, 1983.

196. **Kaibuchi, K., Takai, Y., Sawamura, M., Hoshijima, M., Fujikiwa, T., and Nishizuka, Y.,** Synergistic functions of protein phosphorylation and calcium mobilization in platelet activation, *J. Biol. Chem.,* 258, 6701, 1983.

197. **Rao, A. K., Koike, K., Willis, J., Daniel, J. L., Beckett, C., Hassel, B., Day, H. J., Smith, B. J., and Holmsen, H.,** Platelet secretion defect associated with impaired liberation of arachidonic acid and normal myosin light chain phosphorylation, *Blood,* 64, 914, 1984.

198. **White, J. G.,** Effects of colchicine and vinca alkaloids on human platelets. III. Influence on internal primary contraction and secondary aggregation, *Am. J. Pathol.,* 54, 467, 1969.

199. **Boyle Kay, M. M. and Fudenberg, H. H.,** Inhibition and reversal of platelet activation by cytochalasin B or colemid, *Nature (London),* 244, 288, 1973.

200. **Haslam, R. J., Davidson, M. M. L., and McClenaghan, M. D.,** Cytochalasin B, the blood platelet release reaction and cyclic GMP, *Nature (London),* 253, 455, 1975.

201. **Durham, A. C.,** A unified theory of the control of actin and myosin in nonmuscle movements, *Cell,* 2, 123, 1974.

202. **Bennett, J. P.,** The role of contractile filaments in secretion, *Biochem. Soc. Trans.,* 12, 963, 1984.

203. **Burgoyne, R. D.,** Mechanisms of secretion from adrenal chromaffin cells, *Biochim. Biophys. Acta,* 779, 201, 1984.

204. **Boyles, J. and Bainton, D. F.,** Changes in plasma-membrane-associated filaments during endocytosis and exocytosis in polymorphonuclear leukocytes, *Cell,* 24, 905, 1981.

205. **Howard, T. H. and Oresajo, C. O.,** The kinetics of chemopeptide-induced change in F-actin content, F-actin distribution, and the shape of neutrophils, *J. Cell Biol.,* 101, 1078, 1985.

206. **Sawyer, D. W., Sullivan, J. A., and Mandell, G. L.,** Intracellular free calcium in neutrophils during phagocytosis, *Science,* 230, 663, 1985.

Energy Metabolism During Secretion

Chapter 15

ALTERATIONS IN ENERGY METABOLISM OF SECRETORY CELLS

U. Panten and S. Lenzen

TABLE OF CONTENTS

I. INTRODUCTION

Stimulation of secretion is accompanied by changes in energy metabolism of the secretory cells. These metabolic responses differ considerably in dependence on the type of secretory mechanism. The acid-producing parietal cell is distinguished by a high mitochondrial volume (30 to 40% of the cell volume) and a high rate of basal O_2 consumption (about 15 nmol/min/mg of dry weight), which is markedly enhanced by a strong secretory stimulus.[1-4] The mitochondria of the pancreatic B cell, on the other hand, account for less than 4% of the cell volume and the respiratory rates in resting or active B cells amount only to one third of the corresponding values in parietal cells[5] (Figure 1, Table 1). In view of these differences we confine our article to cells which secrete primarily via typical granular exocytosis. The term secretory cell is used with this restriction. Furthermore, we do not consider cells from very heterogeneous tissues or pathologically transformed cells and we do not discuss long-term changes in energy metabolism, e.g., those caused by cell growth.

II. COMPARTMENTATION OF NUCLEOTIDES

Synthesis, storage, and extrusion of the various secretory products involve energy-requiring processes such as cellular anabolism, membrane recycling, active transport, microtubule assembly, actomyosin contractile activity, phospholipid turnover, cyclase activity, and protein phosphorylation.[6] These reactions are associated with the hydrolysis of high energy phosphates. In addition, nucleoside triphosphates exert allosteric effects on enzymes and ion channels in secretory cells. For instance in pancreatic B cells GTP activates adenylate cyclase and inhibits glutamate dehydrogenase and ATP inhibits a K^+-selective channel of the plasma membrane.[7-10]

The control function of a nucleotide depends on its concentration and efficiency in the microenvironment of its target in the intact cell.[11,12] Most of these local domains are components of the extraorganellar cytoplasm which comprises about half of the cell volume (Table 2). Alterations in the concentration of ATP or other nucleotides in this compartment may be concealed by the content of organelles. This is the case of storage granules in chromaffin cells and platelets which contain high concentrations of various metabolically inert nucleotides.[27-29] However, the low ATP level in β-granules of pancreatic B-cells demonstrates that the storage organelles do not necessarily cause analytical problems in all secretory cells (Table 2). Nucleotides occur in mitochondria and extraorganellar cytoplasm at similar levels (Table 2). Yet, due to the low mitochondrial volume (less than 10% of the cell volume)[1,5,21,23,30-33] the mitochondrial contribution to whole cell data is very small in cells secreting by granular exocytosis. The nucleus constitutes up to 12% of the volume of secretory cells.[5,21,23,31,32] This compartment is freely accessible from the extraorganellar cytoplasm and does not accumulate nucleotides.[34] There is no indication so far that the nucleotide concentrations in the endoplasmic reticulum dissociate from the concentrations in the extraorganellar space.[35] Thus, determination of nucleotides in whole secretory cells reflects the content of the extraorganellar cytoplasm provided that only minor amounts of nucleotides are stored in granules.

Additional compartmentation of the cytoplasm is caused by tight binding of nucleotides to proteins. In resting muscle more than 90% of the extramitochondrial adenosine diphosphate (ADP) is bound, mainly to actin.[36] In platelets up to 50% of the extragranular ADP is associated with actin and does not closely follow changes in the residual extragranular ADP.[37] Though less actin is available for ADP binding in other cells, selective binding of ADP to extraorganellar proteins seems to be a general phenomenon and impedes detection of changes in ADP levels.[38]

Besides membrane barriers and binding, enzyme activities and charge effects represent

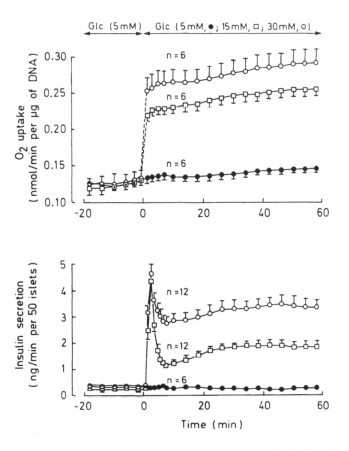

FIGURE 1. Effects of D-glucose (Glc) concentration on the kinetics of insulin secretion and O_2 uptake by pancreatic islets. The experiments were performed at 37°C with islets from fed albino mice exactly as described previously.[105] Values in the curves are means ± SEM of results from n separate experiments.

additional causes for heterogeneity in intracellular distribution of metabolites.[39] The close vicinity of proteins yielding and utilizing nucleoside triphosphates can organize microcompartments for these compounds which equilibrate only incompletely with the bulk phase.[40,41] In local domains with limited access of ATP relative to high consumption, the ATP levels are lower than in the surrounding cytoplasm.[42] Due to charge-charge interaction, the ATP concentration in such microenvironments may be particularly low if the ATP-utilizing enzymes are located in the plasma membrane which carries a net negative charge on its internal surface.[43-45] It has been shown that in intact liver cells the Na^+, K^+-ATPase is exposed to a lower ATP concentration than are extraorganellar enzymes not restricted to the plasma membrane.[46]

In summary, there is considerable evidence for a nonuniform distribution of nucleotides in the extraorganellar space. Thus, measurement of the average nucleotide concentration in this compartment may not always disclose fluctuations in the concentration at sites controlling the function of secretory cells.

III. REGULATION OF HIGH ENERGY PHOSPHATE LEVELS

Activation of exocytosis in secretory cells may coincide with a decrease, with no change, or even with an increase in the levels of ATP and of the phosphorylation potential. The

Table 1
OXYGEN UPTAKE, LACTATE PRODUCTION, AND EXTRAGRANULAR ATP CONCENTRATION IN RESTING SECRETORY CELLS

Cell (species)	Glucose (mM)	O$_2$ uptake	Lactate production	ATP (mM)	Ref.
		(nmol/min per mg of dry weight)			
Platelet (man)	5	1.4[a]	15.1[a]	7.7[a]	51, 52
Mast cell (rat)	5	0.7[b]	1.3[b]	3.5—3.9[b]	53—56
Chromaffin cell (cattle)	10	2.0[c]		1.3[c]	57, 58
Pancreatic exocrine cell (rat, mouse)	10	3.5—4.1[d]	0.25[d]	1.9—2.9[d]	59—61
Pancreatic B cell (mouse)	5	4.8[e]	0.37	5.0[e]	62, 63 Figure 1

Note: Data are for oxygenated medium at 37°C and were calculated using the following values.

[a] 2.8 pg of dry weight and 5.3/μm^3 of extragranular cytoplasm per platelet.[33,64]
[b] 476 pg of dry weight and 357/μm^3 of extragranular space per mast cell.[31,65]
[c] 25% dry weight and 15% extragranular ATP in adrenal medulla.[66,67,89]
[d] Protein content, dry weight, or extragranular cell volume of pancreas comprising 13, 25, or 76.8% of wet weight, respectively.[21,66]
[e] 38.3 ng of DNA and 2.0 nℓ of extragranular cell volume per μg of islet dry weight.[5,68,69]

Table 2
VOLUME[a] AND ATP CONCENTRATION[b] OF SUBCELLULAR COMPARTMENTS[c]

Cell	Extraorganellar cytoplasm	Mitochondria	Secretory granules	Ref.
Cardiac myocyte	58[d] (8.1)[e]	36 (6.0)		13—15
Renal proximal tubular cell	53 (4.3)[e]	30 (2.6)		16—18
Hepatocyte	54 (6.2)[e]	22 (7.5)		19, 20
Pancreatic exocrine cell	54[f]	8.1[f]	6.4[f]	21
Pancreatic B cell	53[g]	3.9[g]	11.5[g] (1)[h]	5, 22, 69
Chromaffin cell		3.4	10.0(150)[i]	23, 24

[a] Values in the table indicate the volumes of the compartments as percent of the cell volume.
[b] The ATP concentrations (mM) in the corresponding compartments are given in parentheses.
[c] Unless stated otherwise the data refer to rat cells.
[d] Myofibrils account for 82% of this value.
[e] Extramitochondrial concentration.
[f] Guinea pig.
[g] Mouse.
[h] Calculated assuming that β-granules contain the same percentage of protein as the total B-cell.[25,26]
[i] Cattle.

outcome depends on the enhancement of fuel catabolism and oxidative phosphorylation relative to augmented utilization of ATP. Glycogenolysis, glycolysis, and oxidative phosphorylation are stimulated by secretagogue-induced events such as increase in cytosolic Ca^{2+} and cyclic nucleotides or decrease in extramitochondrial phosphorylation potential.[47-50] The relative contribution of glycolysis or oxidative phosphorylation to ATP synthesis is quite different in the various secretory cells whereas extragranular ATP concentrations cover a rather narrow range (Table 1).

A. Platelet

When blood platelets are exposed to certain agonists (e.g., thrombin) these cells change shape, aggregate with each other, and secrete the contents of their storage granules.[70] These reactions are energy-requiring processes.[71] The increase in energy consumption accompanying complete functional response is mainly due to secretion from α- and lysosomal granules which costs about seven-fold more energy than dense granule secretion or aggregation.[72-75] Shape change consumes only minor amounts of energy.[75] The work load imposed by these responses lowers the extragranular (or metabolic) ATP level and the adenylate energy charge (AEC = ATP + 0.5 ADP/ATP + ADP + AMP) as demonstrated in platelets labeled by incubation for short periods of time with radioactive adenine or phosphate.[76-78] Under these conditions the adenine nucleotides in the dense granules (the storage or nonmetabolic pool) remain practically unlabeled.[70] Thrombin-induced decrease in extragranular ATP has also been shown by ^{31}P nuclear magnetic resonance.[79] The fall in AEC is partially damped by adenylate deaminase which keeps the AMP level low and induces accumulation of hypoxanthine.[70] Thus, strong platelet responses coincide with a decrease both in ATP and in the size of the adenylate pool.

The fall in metabolic ATP and AEC reflects an imbalance between energy utilization and generation. In human platelets, which lack the creatine-phosphate system,[80] glycolysis is the main rapidly available source for ATP production (Table 1). Lactate formation by platelets incubated in an oxygenated medium supplemented with 5 mM glucose is doubled by a maximally effective thrombin stimulus.[51] The extra lactate formation is in part due to activation of glycogenolysis and lasts much longer than the functional response which is completed within a few minutes.[81]

Unlike glycolysis, mitochondrial O_2 uptake in platelets is enhanced only by a strong stimulus which initiates secretion from lysosomal granules.[51] The rate in mitochondrial respiration is doubled during a 30-sec period and returns to the prestimulatory value after completion of the secretory response.[51] Free fatty acids derived from endogenous lipids are major fuels for the transient burst in O_2 uptake since platelets have a high capacity for fatty acid oxidation.[70] The contribution of mitochondria to platelet energy production may be higher in vivo than in vitro, because normal plasma contains free fatty acids and lactate. The absence of these constituents from the usual incubation media for platelets facilitates lactate efflux and glycolysis and inhibits oxidative phosphorylation.[70]

The process consuming the extra energy supplied by the prolonged increase in glycolytic flux has not been revealed so far. Since significant restoration of the secretory machinery does not occur in platelets.[77] Enhanced ATP production may be used for replenishment of energy stores or for phospholipid synthesis.[75,81]

B. Mast Cell

Histamine release by mast cells is even faster than secretion from platelets. At 37°C the energy-requiring secretory responses to compound 48/80 or to anaphylactic reaction are completed in 10 to 30 sec, respectively.[82] These explosive reactions which release about 50% of the stored histamine coincide with a significant enhancement of ATP utilization in mast cells.[82,83] However, in the presence of exogenous glucose rapid stimulation of glucose metabolism in mast cells prevents decrease of the extragranular ATP level during the secretory responses[55,84] (the ATP content of mast cell granules is negligible[85]). After termination of histamine release, the strong increase in glucose utilization persists for 45 min or more with the various stimuli and probably reflects the energy demand of restorative processes,[82,84] since mast cells can regain their responsiveness.[86,87]

C. Chromaffin Cell

Stimulation of cholinergic receptors in the adrenal medulla triggers catecholamine release which is inhibited by blockade of the energy metabolism in the chromaffin cells.[88] Secretion induced by maximally effective concentrations of acetylcholine or carbachol is accompanied by a 30 to 40% enhancement of oxygen consumption in medullary cells respiring in the presence of glucose.[57,89] Due to the high amounts of ATP in chromaffin granules, the effects of secretagogues on the extragranular ATP content in medullary cells could not be determined so far.[90]

D. Pancreatic Exocrine Cell

The enzymatic secretion induced in the pancreas by proper agonists (e.g., cholinergic agents or cholecystokinin) is dependent on the availability of metabolic energy in the exocrine cells.[59,91] The prolonged secretory responses to high agonist concentrations coincide with a 20% decrease in the ATP content of exocrine pancreas utilizing glucose and amino acids.[59] Since creatine phosphate is present in pancreatic acini in substantial amounts (only 35% less than the ATP content),[61] the fall in phosphorylation potential may be more pronounced than reflected by the fall in ATP (presumably extragranular) and probably represents the major driving force for secretagogue-induced oxygen consumption in the exocrine pancreas. A 10 to 50% increase in respiration and a marked enhancement of amino acid oxidation occur during strong stimulation of enzyme release whereas lactate formation is not changed.[59,92,93] Under normal conditions, glycolysis and glucose oxidation account only for a small proportion of the energy production in the exocrine cells as compared to the contribution of amino acid oxidation.[59,93]

E. Pancreatic B Cell

The pancreatic B cell differs from all other secretory cells in that its major secretagogues exert their primary effects intracellularly, act at millimolar concentrations, and are fuels (e.g., D-glucose, L-leucine, or α-ketoisocaproic acid, KIC) or fuel analogs (e.g., the non-metabolized leucine analog endo-2-aminobicyclo [2.2.1]-heptane-2-carboxylic acid, BCH).[94,95] A close correlation exists between the β-cytotropic activity of different sugars and their rate of metabolism in pancreatic islets.[94,96] This led to the hypothesis that an increase in the catabolism of certain nutrients in the B cell provides signals for insulin release and biosynthesis.[96,97] Since nutrient utilization depends on the rate of ATP turnover which is accelerated during secretion (see aforementioned secretory cells), the proof of the above model requires the identification of reactions involved both in the production of metabolic energy and in the generation of signals for the specific functions of the B cell.

At extracellular concentrations above 5 mM, D-glucose stimulates insulin secretion which is accompanied by enhanced glucose phosphorylation and utilization in pancreatic islets.[94,96] These secretory and metabolic responses reach maximal values at glucose concentrations around 30 mM.[98,99] Above 5 mM D-glucose, more than 80% of glucose utilization occurs via glycolysis which is controlled by B cell glucokinase (apparent K_m value for glucose around 10 mM).[96,98-100] At maximally effective glucose concentrations lactate formation in isolated islets is two- to threefold higher than in the presence of 5 mM D-glucose and accounts for about 50% of glucose utilization.[62,101] Oxidation of glucose amounts to about 30% of its usage and coincides with a small overall inhibition of oxidation of endogenous nutrients

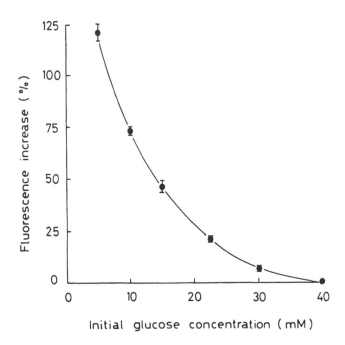

FIGURE 2. Effects of D-glucose concentration on NAD(P)H-fluorescence of single pancreatic islets. The experiments were performed at 37°C with islets from fed albino mice exactly as described previously.[106] After perifusion of islets for 45 min with the indicated glucose concentrations, fluorescence increase was induced by transition to 40 mM glucose and expressed as percentage of the fluorescence decrease (100%) caused by transition to 5 mM glucose 30 min after the first medium change. Values in the curve (except those for 40 mM) are means ± SEM of results from 4 to 7 separate experiments. Fluorescence increase induced by glucose was significant ($p < 0.01$) as tested with the analysis of variance.

in pancreatic islets.[99] Thus, only glucose serves as fuel for the increase in O_2 uptake which accompanies stimulation of insulin release[102,103] (Figure 1).

Respiration is driven by the disequilibrium between the overall reduction potential span across the respiratory chain and the cytosolic phosphorylation potential.[49] Hence, the increased energy demand during enhanced insulin secretion could be satisfied without decrease, or even with an increase in the cytosolic ATP concentration provided that the secretagogue causes sufficient stimulation of hydrogen (reducing equivalents) supply to the respiratory chain of the B cell. Glucose-induced O_2 uptake coincides with a sustained increase in the fluorescence of reduced pyridine nucleotides [NAD(P)H] of pancreatic islets.[104] Both metabolic responses occur with similar dependence on glucose concentration (Figures 1 and 2). Since B cells make up 88% of the volume of mouse islets and contain a low mitochondrial volume,[5] the fluorescence increase mainly represents increase in the content of extramitochondrial NAD(P)H in B cells. However, for the following reasons the fluorimetric responses also reflect reduction of intramitochondrial pyridine nucleotides. First, protein-bound reduced pyridine nucleotides which are detected by the fluorescence technique and represent the major extramitochondrial pools equilibrate rapidly with the free NADH and NADPH.[107] Second, the glycerol phosphate shuttle and the malate aspartate shuttle transfer reducing equivalents into the B cell mitochondria.[108,109] Third, the marked increase in NAD(P)H-fluorescence of mouse islets caused by the mitochondrial substrate α-ketoisocaproic acid indicates rapid exchange of reducing equivalents between the mitochondrial and the extramitochondrial space in the B cell.[110] Thus, the close correlation between glucose-induced respiratory and fluorimetric responses suggests that an increase in the reduction potential span across the respiratory chain is the major driving force for O_2 consumption evoked by insulin-releasing fuels.

The latter view is supported by studies with secretagogues which are metabolized intra-mitochondrially and/or stimulate utilization of intramitochondrial nutrients. These compounds activate glutamate dehydrogenase allosterically (L-leucine, BCH) or serve as partners for intramitochondrial transamination of glutamate and glutamine (phenylpyruvate; KIC; α-ketocaproic acid, KC)[95,111] Thus, they form citrate cycle intermediates and enhance the capacity of the cycle. Increased turnover of the cycle and activation of some reactions outside the cycle augment the hydrogen supply to the respiratory chain of B cells. This is reflected by the kinetics of reduction of extramitochondrial pyridine nucleotides. Increase in NAD(P)H fluorescence of mouse islets exposed to L-leucine, BCH, KIC, or KC is maximal a few minutes after addition of these secretagogues.[110,112] Thereafter the fluorescence traces decline continuously indicating a diminished transfer of reducing equivalents out of the mitochondria, probably due to hydrogen consumption during sustained respiration.

Extracellular concentrations of nutrient secretagogues slightly above their threshold for the initiation of insulin release are sufficient to raise the islet ATP content and adenylate energy charge to maximal values.[97,113] This dissociation between insulin secretion and levels of high energy phosphates does not result from buffering by the creatine-phosphate system because creatine-phosphate was not found in isolated islets.[114] However, the above finding does not rule out changes in the ATP concentration in microcompartments of the extraorganellar space of the B cell. This is concluded from experiments with sulfonylureas (e.g., tolbutamide) which trigger insulin secretion by acting upon specific receptors in the B cell plasma membrane.[115] The secretory profiles typical for these drugs are accompanied by corresponding respiratory profiles[114] (Figure 3). These respiratory responses are evoked mainly by a decrease in the phosphorylation potential, since sulfonylureas do not stimulate catabolic processes significantly and cause only minor changes in the content of NAD(P)H in pancreatic islets[104,116] (Figure 3). Thus, the decrease in the ATP content of islets incubated in the presence of tolbutamide and 5 or 10 mM D-glucose is caused by the energy requirements of the secretory process.[114,116] In conjunction with high glucose concentrations (15 or 30 mM) the expected tolbutamide-induced decrease in islet content of ATP coinciding with the increase in O_2 uptake and insulin secretion is not detected.[114] Hence, the total ATP content of islets which reflects the average concentration in the extraorganellar cytoplasm is not representative for the ATP concentration in microenvironments in which major changes in ATP utilization occur (see section on compartmentation of nucleotides).

The GTP content of isolated pancreatic islets amounts to 36% of the ATP content and is two to three times higher than the GTP content of liver or brain.[117] A high activity of nucleoside diphosphate kinase in islets causes rapid equilibration of the ATP/ADP system with the GTP/GDP system in the extraorganellar cytoplasm.[117] Therefore, glucose-induced augmentation of the ATP level in islets is accompanied by a similar increase in the GTP level.[117]

During insulin release evoked by L-leucine and related compounds or by glucose, 100 or about 90%, respectively, of the extra amount of high energy phosphates produced in isolated islets originate from mitochondrial energy metabolism (see above data for glucose metabolism). In vivo, the contribution of glycolysis is even lower for the same reasons applying to platelets. The respiratory responses reflect stimulation of oxidative phosphorylation, since extramitochondrial O_2 consumption is negligible in pancreatic islets.[103] As sulfonylureas do not enhance insulin synthesis,[115] the correlation between tolbutamide-induced secretory and respiratory profiles favors the view that this drug evokes O_2 uptake by the work load imposed by the release process[114] (Figure 3). However, the biphasic insulin release triggered by glucose or BCH coincides with a monophasic increase in O_2 consumption[105,114] (Figure 1).

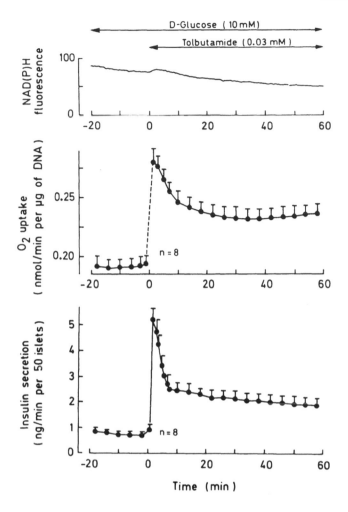

FIGURE 3. Effects of tolbutamide on the kinetics of insulin secretion, O₂ uptake, and NAD(P)H-fluorescence in pancreatic islets. The fluorescence trace is a representative recording from a single islet. At -40 min transition from 5 to 10 m*M* glucose was made and raised the fluorescence from 0 to 100 (arbitrary units). Further details are the same as for Figures 1 and 2.

Furthermore, the half-maximal respiratory response to glucose occurs at a lower concentration than the half maximal secretory response (Figure 1). Finally, O_2 uptake per nanogram of released insulin was higher after elevation of the glucose concentration than after addition of tolbutamide[114] (Figures 1 and 3). These dissociations between secretion and respiration indicate that fuel-induced O_2 consumption results not only from insulin release but also from the energy demand of augmented biosynthetic and restorative processes.

IV. CONTROL OF FUNCTION BY HIGH ENERGY PHOSPHATES

Platelets containing a low level of ATP in their extragranular space function normally provided that their adenylate energy charge remains unchanged.[70,75] In platelets depleted of substrate a close correlation between adenylate energy charge and secretory response exists.[118] These findings suggest that cellular functions depend on the free enthalpy for ATP hydrolysis at certain enzymes catalyzing energy consuming reactions.

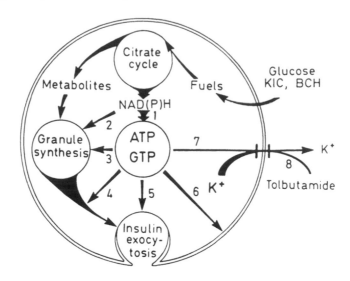

FIGURE 4.   Diagram illustrating interactions between energy metabolism and function of the pancreatic B cell. (1) Oxidative phosphorylation; (2 and 3) utilization of NADPH, ATP, and GTP for granule biosynthesis; (4) utilization of ATP and GTP by motile events; (5) stimulation of exocytosis by GTP; (6) utilization of ATP and GTP by membrane function; (7 and 8) blockade of potassium channel by ATP or tolbutamide.

In contrast, insulin release is diminished whenever the islet content of ATP falls below its maximal value. This is the case at low concentrations of nutrient secretagogues.[97,113] Moreover, sulfonylurea-induced work load causes a decrease in the ATP content of islets which is accompanied by a typical decline in the secretory profile[114] (Figure 3). These observations are consistent with the proposal that a K^+ channel which is blocked by ATP mediates the depolarization of the B cell plasma membrane initiated by insulin-releasing fuels [9,10,119] (Figure 4). In support of this view, sulfonylureas inhibit the same K^+ channel by direct interaction[120] (Figure 4). Depolarization opens voltage-dependent Ca^{2+} channels and the resulting elevation of the cytosolic Ca^{2+} concentration triggers insulin secretion. The above K^+ channel is regulated by the ATP concentration in its immediate vicinity which is not reflected by the average concentration in islets for reasons discussed in Sections II and III.E.

However, additional targets for high energy phosphates may be involved in the control of B cell function (Figure 4). GTP is required for receptor coupling to a K^+ channel in cardiac cells and activates exocytosis in mast cells directly.[121,122] Furthermore, the suggestion that GTP represents a central regulator of cellular anabolism implies stimulation of biosynthesis in the pancreatic B cell by GTP.[117,123]

The concentrations of high energy phosphates and the phosphorylation potential in certain local domains of the B cell cytoplasm depend on the hydrogen supply to the respiratory chain. The citrate cycle is the major intramitochondrial source of reducing equivalents. Its limited capacity in the resting B cell is not enlarged significantly by indirect mechanisms, e.g., activation of mitochondrial dehydrogenases due to increase in matrix Ca^{2+}.[50] This is indicated by the failure of tolbutamide to enhance catabolic processes and the NAD(P)H content in pancreatic islets[104,116] (Figure 3). However, insulin releasing fuels and their analogs furnish intermediates for the citrate cycle and directly augment the availability of substrate for the respiratory chain.[95,105] Hence, the pancreatic B cell is well-equipped for use of its energy metabolism to recognize β-cytotropic fuels.

ACKNOWLEDGMENT

The original work described here was supported by the Deutsche Forschungsgemeinschaft.

REFERENCES

1. **Berglindh, T.,** The mammalian gastric parietal cell in vitro, *Annu. Rev. Physiol.,* 46, 377, 1984.
2. **Croft, D. N. and Ingelfinger, F. J.,** Isolated gastric parietal cells: oxygen consumption, electrolyte content and intracellular pH, *Clin. Sci.,* 37, 491, 1969.
3. **Soll, A. H.,** The actions of secretagogues on oxygen uptake by isolated mammalian parietal cells, *J. Clin. Invest.,* 61, 370, 1978.
4. **Sachs, G., Spenney, J. G., and Lewin, M.,** H^+ transport: regulation and mechanism in gastric mucosa and membrane vesicles, *Physiol. Rev.,* 58, 106, 1978.
5. **Dean, P. M.,** Ultrastructural morphometry of the pancreatic β-cell, *Diabetologia,* 9, 115, 1973.
6. **Akkerman, J. W. N.,** The energy requirement of secretion, in *Energetics of Secretion Responses,* Vol. 2, Akkerman, J. W. N., Ed., CRC Press, Boca Raton, Fla., 1988, chap. 16.
7. **Montague, W. and Howell, S. L.,** Cyclic AMP and the physiology of the islets of Langerhans, *Adv. Cyclic Nucleotide Res.,* 6, 201, 1975.
8. **Sener, A., Owen, F., Malaisse-Lagae, F., and Malaisse, W.,** The stimulus-secretion coupling of amino acid-induced insulin release. XI. Kinetics of deamination and transamination reactions, *Hormone Metab. Res.,* 14, 405, 1982.
9. **Cook, D. L. and Hales, C. N.,** Intracellular ATP directly blocks K^+ channels in pancreatic B-cells, *Nature (London),* 311, 271, 1984.
10. **Ashcroft, F. M., Harrison, D. E., and Ashcroft, S. J. H.,** Glucose induces closure of single potassium channels in isolated rat pancreatic β-cells, *Nature (London),* 312, 446, 1984.
11. **Masters, C. J.,** Metabolic control and the microenvironment, *Curr. Top. Cell. Regul.,* 12, 75, 1977.
12. **Soltoff, S. P. and Mandel, L. J.,** Active ion transport in the renal proximal tubule. III. The ATP dependence of the Na pump, *J. Gen. Physiol.,* 84, 643, 1984.
13. **Page, E. and McCallister, L. P.,** Quantitative electron microscopic description of heart muscle cells. Application to normal, hypertrophied and thyroxin-stimulated hearts, *Am. J. Cardiol.,* 31, 172, 1973.
14. **Page, E., Early, J., and Power, P.,** Normal growth of ultrastructures in rat left ventricular myocardial cells, *Circ. Res.,* 34 + 35(Suppl. II), 17, 1974.
15. **Kauppinen, R. A., Hiltunen, J. K., and Hassinen, I. E.,** Subcellular distribution of phosphagens in isolated perfused rat heart, *FEBS Lett.,* 112, 273, 1980.
16. **Orsoni, J., Rohr, H. P., and Gloor, F.,** Morphometric characterization of the different segments of the renal tubular apparatus in the rat, *Pathol. Eur.,* 4, 345, 1969.
17. **Pfaller, W. and Rittinger, M.,** Quantitative morphology of the rat kidney, *Int. J. Biochem.,* 12, 17, 1980.
18. **Pfaller, W., Guder, W. G., Gstraunthaler, G., Kotanko, P., Jehart, I., and Pürschel, S.,** Compartmentation of ATP within renal proximal tubular cells, *Biochim. Biophys. Acta,* 85, 152, 1984.
19. **Weibel, F., Stäubli, W., Gnägi, H. R., and Hess, F. A.,** Correlated morphometric and biochemical studies on the liver cell, *J. Cell Biol.,* 42, 68, 1969.
20. **Schwenke, W. D., Soboll, S., Seitz, H. J., and Sies, H.,** Mitochondrial and cytosolic ATP/ADP ratios in rat liver in vivo, *Biochem. J.,* 200, 405, 1981.
21. **Bolender, R. P.,** Stereological analysis of the guinea pig pancreas, *J. Cell Biol.,* 61, 269, 1974.
22. **Leitner, J. W., Sussman, K. E., Vatter, A. E., and Schneider, F. H.,** Adenine nucleotide in the secretory granule fraction of rat islets, *Endocrinology,* 95, 662, 1975.
23. **Nordmann, J. J.,** Combined stereological and biochemical analysis of storage and release of catecholamines in the adrenal medulla of the rat, *J. Neurochem.,* 42, 434, 1983.
24. **Philipps, J. H., Allison, Y. P., and Morris, S. J.,** The distribution of calcium, magnesium, copper and iron in the bovine adrenal medulla, *Neuroscience,* 2, 147, 1977.
25. **Howell, S. L. and Tyhurst, M.,** The insulin storage granule, in *The Secretory Granule,* Poisner, A. M. and Trifaro, J. M., Eds., Elsevier, Amsterdam, 1982, 155.
26. **Hutton, J. C.,** Secretory granules, *Experientia,* 40, 1025, 1984.
27. **Winkler, H. and Carmichael, S. W.,** The chromaffin granule, in *The Secretory Granule,* Poisner, A. M. and Trifaró, J. M., Eds., Elsevier, Amsterdam, 1982, 3.
28. **Holmsen, H., Day, H. J., and Storm, E.,** Adenine nucleotide metabolism of blood platelets. VI. Subcellular localization of nucleotide pools with different functions in the platelet release reaction, *Biochim. Biophys. Acta,* 186, 254, 1969.

29. **Holmsen, H. and Weiss, H. J.**, Secretable storage pools in platelets, *Annu. Rev. Med.*, 30, 119, 1979.

30. **Altenähr, E. and Leonhardt, F.**, Suppression of parathyroid gland acivity by magnesium, *Virchows Arch. Abt. A Pathol. Anat.*, 355, 297, 1972.

31. **Helander, H. and Bloom, G. D.**, Quantitative analysis of mast cell structure, *J. Microsc.*, 100, 315, 1974.

32. **Cope, G. H. and Williams, M. A.**, Exocrine secretion in the parotid gland: a stereological analysis at the electron microscopic level of the zymogen granule content before and after isoprenaline-induced degranulation, *J. Anat.*, 116, 269, 1973.

33. **Stahl, K., Themann, H., and Dame, W. R.**, Ultrastructural morphometric investigations on normal human platelets, *Haemostasis*, 7, 242, 1978.

34. **Siebert, G.**, The biochemical environment of the mammalian nucleus, *Sub-Cell. Biochem.*, 1, 277, 1972.

35. **Elbers, R., Heldt, H. W., Schmucker, P., Soboll, S., and Wiese, H.**, Measurement of the ATP/ADP ratio in mitochondria and in the extramitochondrial compartment by fractionation of freeze-stopped liver tissue in non-aqueous media, *Hoppe-Seyler's Z. Physiol. Chem.*, 355, 378, 1974.

36. **Hebisch, S., Soboll, S., Schwenen, M., and Sies, H.**, Compartmentation of high-energy phosphates in resting and working rat skeletal muscle, *Biochim. Biophys. Acta*, 764, 117, 1984.

37. **Daniel, J. L., Robkin, L., Molish, I. R., and Holmsen, H.**, Determination of the ADP concentration available to participate in energy metabolism in an actin-rich cell, the platelet, *J. Biol. Chem.*, 254, 7870, 1979.

38. **Gankema, H. S., Groen, A. K., Wanders, R. J. A., and Tager, J. M.**, Measurement of binding of adenine nucleotides and phosphate to cytosolic proteins in permeabilized rat-liver cells, *Eur. J. Biochem.*, 131, 447, 1983.

39. **Srere, P. A. and Mosbach, K.**, Metabolic compartmentation: symbiotic, organellar, multienzymic, and microenvironmental, *Annu. Rev. Microbiol.*, 28, 61, 1974.

40. **Moyer, J. D. and Henderson, J. F.**, Compartmentation of intracellular nucleotides in mammalian cells, *Crit. Rev. Biochem.*, 19, 45, 1985.

41. **Saks, V. A., Kuznetsov, A. V., Kupriyanov, V. V., Miceli, M. V., and Jacobus, W. E.**, Creatine kinase of rat heart mitochondria. The demonstration of functional coupling to oxidative phosphorylation in an inner membrane-matrix preparation, *J. Biol. Chem.*, 260, 7757, 1985.

42. **Opie, L. H.**, High energy phosphate compounds, in *Cardiac Metabolism*, Drake-Holland, A. J. and Noble, M. I. M., Eds., John Wiley & Sons, New York, 1983, 279.

43. **Hornby, W. E. and Lilly, M. D.**, Some changes in the reactivity of enzymes resulting from their chemical attachment to water-insoluble derivatives of cellulose, *Biochem. J.*, 107, 669, 1968.

44. **DeSimone, J. A.**, Perturbations in the structure of the double layer at an enzymic surface, *J. Theor. Biol.*, 68, 225, 1977.

45. **Honig, B. H., Hubbell, W. L., and Flewelling, R. E.**, Electrostatic interactions in membranes and proteins, *Annu. Rev. Biophys. Biophys. Chem.*, 15, 163, 1986.

46. **Aw, T. Y. and Jones, D. P.**, ATP concentration gradients in cytosol of liver cells during hypoxia, *Am. J. Physiol.*, 249, C385, 1985.

47. **Hers, H. G. and Hue, L.**, Gluconeogenesis and related aspects of glycolysis, *Annu. Rev. Biochem.*, 52, 617, 1983.

48. **Ramaiah, A.**, Pasteur effect and phosphofructokinase, *Curr. Top. Cell. Regul.*, 8, 297, 1974.

49. **Hansford, R. G.**, Control of mitochondrial substrate oxidation, *Curr. Top. Bioenerget.*, 10, 217, 1980.

50. **Hansford, R. G.**, Relation between mitochondrial calcium transport and control of energy metabolism, *Rev. Physiol. Biochem. Pharmacol.*, 102, 1, 1985.

51. **Akkerman, J. W. and Holmsen, H.**, Interrelationships among platelet responses: studies on the burst in proton liberation, lactate production, and oxygen uptake during platelet aggregation and Ca^{2+} secretion, *Blood*, 57, 956, 1981.

52. **Akkerman, J. W., Nieuwenhuis, H. K., Mommersteeg-Leautaud, M., Gorter, G., and Sixma, J. J.**, ATP-ADP compartmentation in storage pool deficient platelets: correlation between granule-bound ADP and bleeding time, *Br. J. Haematol.*, 55, 135, 1983.

53. **Chakravarty, N. and Zeuthen, E.**, Respiration of rat peritoneal mast cells, *J. Cell Biol.*, 25, 113, 1965.

54. **Peterson, C.**, Histamine release induced by compound 48/80 from isolated rat mast cells: dependence on endogenous ATP, *Acta Pharmacol. Toxicol.*, 34, 356, 1974.

55. **Diamant, B., Norn, S., Felding, P., Olsen, N., Ziebell, A., and Nissen, J.**, ATP level and CO_2 production of mast cells in anaphylaxis, *Int. Arch. Allergy*, 47, 894, 1974.

56. **Johansen, T. and Chakravarty, N.**, The utilization of adenosine triphosphate in rat mast cells during histamine release induced by anaphylactic reaction and compound 48/80, *Naunyn-Schmiedeberg's Arch. Pharmacol.*, 288, 243, 1975.

57. **Bevington, A. and Radda, G. K.**, Enhanced oxygen consumption in adrenal medulla on stimulation with acetylcholine, *Biochem. Pharmacol.*, 34, 211, 1985.

58. **Rojas, E., Pollard, H. B., and Heldman, E.,** Real-time measurements of acetylcholine-induced release of ATP from bovine medullary chromaffin cells, *FEBS Lett.*, 185, 323, 1985.
59. **Bauduin, H., Colin, M., and Dumont, J. E.,** Energy sources for protein synthesis and enzymatic secretion in rat pancreas in vitro, *Biochim. Biophys. Acta*, 174, 722, 1969.
60. **Schulz, I., Heil, K., Kribben, A., Sachs, G., and Haase, W.,** Isolation and functional characterization of cells from the exocrine pancreas, in *Biology of Normal and Cancerous Exocrine Pancreatic Cells*, Ribet, A., Pradayrol, L., and Susini, C., Eds., Elsevier, Amsterdam, 1980, 3.
61. **Matschinsky, F.,** Enzymes, metabolites, and cofactors involved in intermediary metabolism of islets of Langerhans, in *Handbook of Physiology*, Section 7, Vol. 1, Steiner, D. F. and Freinkel, N., Eds., American Physiological Society, Washington, D.C., 1972, 199.
62. **Ashcroft, S. J. H., Hedeskov, C. J., and Randle, P. J.,** Glucose metabolism in mouse pancreatic islets, *Biochem. J.*, 118, 143, 1970.
63. **Hellman, B., Idahl, L.-A., and Danielsson, A.,** Adenosine triphosphate levels of mammalian pancreatic B cells after stimulation with glucose and hypoglycemic sulfonylureas, *Diabetes*, 18, 509, 1969.
64. **Marcus, A. J. and Zucker, M. B.,** *The Physiology of Blood Platelets*, Grune & Stratton, New York, 1965, 1.
65. **Diamant, B. and Lowry, O. H.,** Dry weight determination of single lyophilized mast cells of the rat, *J. Histochem. Cytochem.*, 14, 519, 1966.
66. **Panten, U.,** unpublished data for rat tissues, 1986.
67. **Ungar, A. and Phillips, J. H.,** Regulation of the adrenal medulla, *Physiol. Rev.*, 63, 787, 1983.
68. **Beckmann, J., Holze, S., Lenzen, S., and Panten, U.,** Quantification of cells in islets of Langerhans using DNA determination, *Acta Diabetol. Lat.*, 18, 51, 1981.
69. **Hellman, B., Sehlin, J., and Täljedal, I. B.,** Transport of α-aminoisobutyric acid in mammalian pancreatic β-cells, *Diabetologia*, 7, 256, 1971.
70. **Holmsen, H., Salganicoff, L., and Fukami, M. H.,** Platelet behaviour and biochemistry, in *Haemostasis: Biochemistry, Physiology, and Pathology*, Ogston, D. and Bennett, B., Eds., John Wiley & Sons, New York, 1977, 239.
71. **Mürer, E. H.,** Release reaction and energy metabolism in blood platelets with special reference to the burst in oxygen uptake, *Biochim. Biophys. Acta*, 162, 320, 1968.
72. **Akkerman, J. W. N., Gorter, G., Schrama, L., and Holmsen, H.,** A novel technique for rapid determination of energy consumption in platelets. Demonstration of different energy consumption associated with three secretory responses, *Biochem. J.*, 210, 145, 1983.
73. **Verhoeven, A. J. M., Mommersteeg, M. E., and Akkerman, J. W. N.,** Quantification of energy consumption in platelets during thrombin-induced aggregation and secretion. Tight coupling between platelet responses and the increment in energy consumption, *Biochem. J.*, 221, 777, 1984.
74. **Verhoeven, A. J. M., Gorter, G., Mommersteeg, M. E., and Akkerman, J. W.,** The energetics of early platelet responses. Energy consumption during shape change and aggregation with special reference to protein phosphorylation and polyphosphoinositide cycle, *Biochem. J.*, 228, 451, 1985.
75. **Verhoeven, A. J. M.,** Energetics of platelet responses, M.D. thesis, University of Utrecht, Utrecht, The Netherlands, 1985.
76. **Mills, D. C. B.,** Changes in the adenylate energy charge in human blood platelets induced by adenosine diphosphate, *Nature (London)*, 24, 220, 1973.
77. **Holmsen, H.,** Biochemistry of the platelet release reaction, in *Biochemistry and Pharmacology of Platelets*, Ciba Foundation Symp. 35, Elsevier, Amsterdam, 1975, 175.
78. **Holmsen, H., Dangelmaier, C. A., and Akkerman, J. W. N.,** Determination of levels of glycolytic intermediates and nucleotides in platelets by pulse-labeling with ^{32}P orthophosphate, *Anal. Biochem.*, 131, 266, 1983.
79. **Ugurbil, K., Holmsen, H., and Schulman, R. G.,** Adenine nucleotide storage and secretion in platelets as studied by ^{31}P nuclear magnetic resonance, *Proc. Natl. Acad. Sci. U.S.A.*, 76, 2227, 1979.
80. **Meltzer, H. Y. and Guschwan, A.,** Type I (brain type) creatine phosphokinase (CPK) activity in rat platelets, *Life Sci.*, 11, 121, 1972.
81. **Akkerman, J. W. N.,** Regulation of carbohydrate metabolism in platelets. A review, *Thrombosis Haemostas.*, 39, 712, 1978.
82. **Chakravarty, N.,** Metabolic changes in mast cells associated with histamine release, in *Handbook of Experimental Pharmacology*, Vol. 18, Part 2, Rocha e Silva, M., Ed., Springer-Verlag, Berlin, 1978, 93.
83. **Johansen, T.,** Utilization of adenosine triphosphate in rat mast cells during and after secretion of histamine in response to compound 48/80, *Acta Pharmacol. Toxicol.*, 53, 245, 1983.
84. **Svendstrup, F. and Chakravarty, N.,** Glucose metabolism in rat mast cells during histamine release, *Exp. Cell Res.*, 106, 223, 1977.
85. **Uvnäs, B.,** The mechanism of histamine release from mast cells, in *Handbook of Experimental Pharmacology*, Vol. 18, Part 2, Rocha e Silva, M., Ed., Springer-Verlag Berlin, 1978, 75.

86. **Thon, I. L. and Uvnäs, B.**, Degranulation and histamine release, two consecutive steps in the response of rat mast cells to compound 48/80, *Acta Physiol. Scand.*, 71, 303, 1967.

87. **Holm Nielsen, E., Bytzer, P., Clausen, J., and Chakravarty, N.**, Electron microscopic study of the regeneration in vitro of rat peritoneal mast cells after histamine secretion, *Cell Tissue Res.*, 216, 635, 1981.

88. **Viveros, O. H.**, Mechanism of secretion of catecholamines from adrenal medulla, in *Handbook of Physiology*, Section 7, Vol. 6, Blaschko, H., Sayers, G., and Smith, A. D., Eds. American Physiological Society, Washington, D.C., 1975, 389.

89. **Banks, P.**, Effects of stimulation by carbachol on the metabolism of the bovine adrenal medulla, *Biochem. J.*, 97, 555, 1965.

90. **Bevington, A., Briggs, R. W., Radda, G. K., and Thulborn, K. R.**, Phosphorus-31 nuclear magnetic resonance studies of pig adrenal glands, *Neuroscience*, 11, 281, 1984.

91. **Hokin, L. E.**, Metabolic aspects and energetics of pancreatic secretion, in *Handbook of Physiology*, Section 6, Vol. 2, Code, C. F., Ed., American Physiological Society, Washington, D.C., 1967, 935.

92. **Dickman, S. R. and Morrill, G. A.**, Stimulation of respiration and secretion of mouse pancreas in vitro, *Am. J. Physiol.*, 190, 403, 1957.

93. **Danielsson, Å. and Sehlin, J.**, Transport and oxidation of amino acids and glucose in the isolated exocrine mouse pancreas: effects of insulin and pancreozymin, *Acta Physiol. Scand.*, 91, 557, 1974.

94. **Hedeskov, C. J.**, Mechanism of glucose-induced insulin secretion, *Physiol. Rev.*, 60, 442, 1980.

95. **Panten, U., Zielmann, S., Joost, H.-G., and Lenzen, S.**, Branched chain amino and keto acids — tools for the investigation of fuel recognition mechanism in pancreatic B-cells, in *Branched Chain Amino and Keto Acids in Health and Disease*, Adibi, S. A., Fekl, W., Langenbeck, U., and Schauder, P., Eds., S. Karger, Basel, 1984, 134.

96. **Ashcroft, S. J. H.**, Glucoreceptor mechanisms and the control of insulin release and biosynthesis, *Diabetologia*, 18, 5, 1980.

97. **Malaisse, W. J., Sener, A., Herchuelz, A., and Hutton, J. C.**, Insulin release: the fuel hypothesis, *Metabolism*, 28, 373, 1979.

98. **Meglasson, M. D. and Matschinsky, F. M.**, New perspectives on pancreatic islet glucokinase, *Am. J. Physiol.*, 246, E1, 1984.

99. **Sener, A. and Malaisse, W. J.**, Nutrient metabolism in islet cells, *Experientia*, 40, 1026, 1984.

100. **Lenzen, S., Tiedge, M., and Panten, U.**, Glucokinase in pancreatic B-cells and its inhibition by alloxan, *Acta Endocrinol. (Kobh.)*, 115, 21, 1987.

101. **Sener, A. and Malaisse, W. J.**, Measurement of lactic acid in nanomolar amounts. Reliability of such a method as an index of glycolysis in pancreatic islets, *Biochem. Med.*, 15, 34, 1976.

102. **Hellerström, C.**, Effects of carbohydrates on the oxygen consumption of isolated pancreatic islets of mice, *Endocrinology*, 81, 105, 1967.

103. **Hutton, J. C. and Malaisse, W. J.**, Dynamics of O_2 consumption in rat pancreatic islets, *Diabetologia*, 18, 395, 1980.

104. **Panten, U., Christians, J., v. Kriegstein, E., Poser, W., and Hasselblatt, A.**, Effect of carbohydrates upon fluorescence of reduced pyridine nucleotides from perifused isolated pancreatic islets, *Diabetologia*, 9, 477, 1973.

105. **Panten, U., Zielmann, S., Langer, J., Zünkler, B.-J., and Lenzen, S.**, Regulation of insulin secretion by energy metabolism in pancratic B-cell mitochondria. Studies with a non-metabolizable leucine analogue, *Biochem. J.*, 219, 189, 1984.

106. **Panten, U., Ishida, H., Schauder, P., Frerichs, H., and Hasselblatt, A.**, A versatile microperifusion system, *Anal. Biochem.*, 82, 317, 1977.

107. **Sies, H.**, Nicotinamide nucleotide compartmentation, in *Metabolic Compartmentation*, Sies, H., Ed., Academic Press, New York, 1982, 205.

108. **Mac Donald, M. J.**, High content of mitochondrial glycerol-3-phosphate dehydrogenase in pancreatic islets and its inhibition by diazoxide, *J. Biol. Chem.*, 256, 8287, 1981.

109. **Mac Donald, M. J.**, Evidence for malate aspartate shuttle in pancreatic islets, *Arch. Biochem. Biophys.*, 213, 643, 1982.

110. **Panten, U.**, Effects of alpha-ketomonocarboxylic acids upon insulin secretion and metabolism of isolated pancreatic islets, *Naunyn-Schmiedeberg's Arch. Pharmacol.*, 291, 405, 1975.

111. **Lenzen, S., Schmidt, W., and Panten, U.**, Transamination of neutral amino acids and 2-keto acids in pancreatic B-cell mitochondria, *J. Biol. Chem.*, 260, 12629, 1985.

112. **Panten, U. and Christians, J.**, Effects of 2-endoamino-norbornane-2-carboxylic acid upon insulin secretion and fluorescence of reduced pyridine nucleotudes of isolated perifused pancreatic islets, *Naunyn-Schmiedeberg's Arch. Pharmacol.*, 276, 55, 1973.

113. **Ashcroft, S. J. H., Weerasinghe, L. C. C., and Randle, P.**, Interrelationship of islet metabolism, adenosine triphosphate content and insulin release, *Biochem. J.*, 132, 223, 1973.

114. **Panten, U., Zünkler, B.-J., Scheit, S., Kirchhoff, K., and Lenzen, S.**, Regulation of energy metabolism in pancreatic islets by glucose and tolbutamide, *Diabetologia*, 29, 648, 1987.

115. **Hellman, B. and Täljedal, I.-B.,** Effects of sulfonylurea derivatives on pancreatic β-cells, in *Handbook of Experimental Pharmacology,* Vol. 32, Part 2, Hasselblatt, A. and von Bruchhausen, F., Eds., Springer-Verlag, Berlin, 1975, 175.
116. **Kawazu, S., Sener, A., Couturier, E., and Malaisse, W. J.,** Metabolic, cationic and secretory effects of hypoglycemic sulfonylureas in pancreatic islets, *Naunyn-Schmiedeberg's Arch. Pharmacol.,* 312, 277, 1980.
117. **Zünkler, B.-J., Lenzen, S., and Panten, U.,** D-Glucose enhances GTP content in mouse pancreatic islets, *IRCS Med. Sci.,* 14, 354, 1986.
118. **Akkerman, J. W. N., Gorter, G., Soons, H., and Holmsen, H.,** Close correlation between platelet responses and adenylate energy charge during transient substrate depletion, *Biochim. Biophys. Acta,* 760, 34, 1983.
119. **Rorsman, P. and Trube, G.,** Glucose dependent K^+-channels in pancreatic β-cells are regulated by intracellular ATP, *Pflugers Arch.,* 405, 305, 1985.
120. **Sturgess, N. C., Ashford, M. L. J., Cook, D. L., and Hales, C. N.,** The sulfonylurea receptor may be an ATP-sensitive potassium channel, *Lancet,* ii, 474, 1985.
121. **Pfaffinger, P. J., Martin, J. M., Hunter, D. D., Nathanson, N. M., and Hille, B.,** GTP-binding proteins couple cardiac muscarinic receptors to a K channel, *Nature (London),* 317, 536, 1985.
122. **Fernandez, J. M., Neher, E., and Gomperts, B. D.,** Capacitance measurements reveal stepwise fusion events in degranulating mast cells, *Nature (London),* 312, 453, 1984.
123. **Pall, M. A.,** GTP: a central regulator of cellular anabolism, *Curr. Top. Cell. Regul.,* 25, 1, 1985.

Chapter 16

THE ENERGY REQUIREMENT OF SECRETION — A STUDY ON BLOOD PLATELETS

Jan-Willem N. Akkerman

TABLE OF CONTENTS

I. INTRODUCTION

Human blood platelets are small, anucleated cells synthesized by megakaryocytes in the bone marrow. Under normal conditions they circulate in the blood stream in a dormant state but once activated the cells rapidly undergo a change in shape and form pseudopods. The contents of different types of secretory granules are liberated and the cells become sticky enabling the formation of aggregates that prevent excessive blood loss and in pathological conditions form thrombi that may obstruct the vessels. Platelets are activated by a variety of agonists for which they possess specific receptors on their plasma membrane. Among the best studied are collagen fibers, which are abundantly present in the vessel wall and become exposed when a vessel is disrupted; thrombin, which is a key factor in the coagulation cascade; ADP which is liberated from other platelets; and platelet activating factor (PAF), a phospholipid synthesized by macrophages, endothelial cells, and platelets. The sequence of responses initiated by these agonists generally involves: (1) a change in cell shape leading to a disk to sphere transformation, (2) aggregation, and (3) secretion of different types of secretory granules. The velocity and extent with which these responses take place vary greatly among different agonists. For example, thrombin, known as a strong agonist, activates this cascade within a few seconds and triggers almost complete degranulation whereas ADP, a weak stimulator, induces slow responses and only partial liberation of granule contents. Much depends on whether an agonist can activate two mechanisms that enhance aggregation and secretion. One mechanism is the formation of prostaglandin metabolites, specifically the endoperoxides PGG_2 and PGH_2 and thromboxane A_2. These prostanoids activate the cells via binding to intracellular receptors. A second is the liberation of ADP into the extracellular medium from secretion granules, which binds to ADP-receptors on the plasma membrane thereby activating the cell. Each of these metabolites serves in positive feedback loops thereby helping to complete aggregation and secretion within a few minutes.

Much insight in the mechanism underlying aggregation and secretion has been gained from patients suffering from platelet abnormalities in membrane components, signal processing, or secretion characteristics. Artificially, inhibition of platelet responses can be obtained with a variety of agents that interfere with the various steps in the secretion process. Among these, metabolic inhibitors are most effective indicating a major role for metabolic energy in the induction and execution of secretion by platelets.

II. ENERGY METABOLISM IN UNSTIMULATED PLATELETS

A. ATP-Resynthesizing Pathways

The main source for the generation of rapidly accessible energy in platelets is glucose, which is taken up from the plasma and converted to lactate and CO_2 in glycolysis and mitochondrial respiration, respectively.[1] Under normal conditions (> 1 mM glucose, saturated O_2) anaerobic and aerobic glucose catabolism contribute almost equally to ATP resynthesis together providing sufficient energy to satisfy the various ATP-consuming processes in the cell.[1,2] Although abundantly present in platelets, glycogen is not consumed unless lack of extracellular glucose requires additional carbohydrate catabolism.[3] Apart from carbohydrates the mitochondria use lipids as a source for oxidative ATP generation. The energy liberated in glycolysis and oxidative phosphorylation is used for resynthesis of ATP. Unlike platelets from birds and other species, human platelets lack the creatine-phosphate creatine-phosphokinase system.[4] Furthermore, they contain only minor quantities of GTP and UTP making ATP the major intermediate that transfers energy from sites of generation to sites of utilization (Figure 1). Since platelets lack the key enzymes required for gluco- and glyconeogenesis and ATP-consuming futile cycles are probably absent, the rate of ATP resynthesis can be calculated from the rates of lactate accumulation and mitochondrial O_2

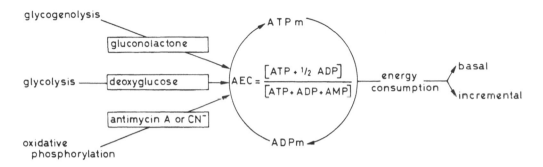

FIGURE 1. Energy production and consumption in human platelets. Platelets produce rapidly accessible energy in glycolysis, glycogenolysis, and oxidative phosphorylation. The energy is stored in the phosphorylation of metabolic ADP to ATP. Metabolic ATP-ADP (ATPm-ADPm) are the major transducers of energy from sites of generation to sites of utilization, since human platelets lack the creatine phosphate-creatine phosphokinase system. The energy stored in metabolic ATP is utilized in the different energy consuming processes of the unstimulated platelet, the so-called basal energy consumption. When platelets are stimulated to shape change, aggregation and secretion extra energy is consumed (incremental energy consumption).

uptake or CO_2 formation.[5] Unstimulated platelets produce 2 to 3 μmol lactate and consume 0.5 to 0.7 μ at O^- in the mitochondria per minute per 10^{11} platelets. Together they resynthesize ATP at a rate of 3 to 6 μmol ATP equivalents · min^{-1} · 10^{-11} cells.* The glycolytic flux is controlled by the entry of glucose and the activities of the enzymes hexokinase, 6-phosphofructokinase, and aldolase. Under normal conditions the glucose transport system is far displaced from equilibrium which together with its location at the beginning of a long metabolic route makes it a prime site for flux control. When mitochondrial respiration is inhibited, the glycolytic flux increases (Pasteur effect) and becomes directly dependent on the extracellular glucose concentration in the range between 10 and 1000 μM.[3] At higher concentrations the uptake is slowed down and adapted to the energy requirement of the cell. The regulatory role of hexokinase is difficult to evaluate since it is almost completely bound to the outer mitochondrial membrane. Apart from its role in the glycolytic pathway, it may play a role as an ADP translocator thereby forming an important link between anaerobic and aerobic energy generation.[5] 6-Phosphofructokinase has been purified and extensively studied. The enzyme shows cooperative kinetics towards its substrate fructose-6-phosphate, a property that can be enhanced or depressed by a variety of metabolites such as citrate, AMP, H^+-ions etc. Its other substrate, $MgATP^{2-}$, exhibits cooperative inhibition at higher concentrations, a process that is affected by the free constituents of the $MgATP^{2-}$ complex. Activation of the glycolytic flux is accompanied with a decrease in fructose-6-phosphate and an extensive accumulation of fructose 1,6-bisphosphate, clearly illustrating a rapid change in phosphofructokinase activity. Similar changes are also seen for dihydroxyacetone-phosphate suggesting that aldolase also contributes to flux control. The control of mitochondrial ATP resynthesis in platelets is poorly understood. Measurements of citric acid cycle intermediates point to a control function by fumerase.[5]

B. Regulation of ATP Homeostasis

In unstimulated platelets the steady-state level of ATP is kept constant when glucose catabolism to lactate and mitochondrial degradation of sugars and lipids provide an ATP resynthesis of 3 to 6 μmol ATPeq · min^{-1} · 10^{-11} platelets, assuming a 1:1 and 1:3 stoichiometry with ATP resynthesis for lactate production and mitochondrial O^- uptake, respectively.[5,6] Any increase in energy need is met by a transient fall in ATP followed by

---

* ATP equivalent (ATPeq) represents the energy liberated in the conversion of ATP to ADP. Thus, ATP and ADP represent 2 and 1 ATPeq, respectively.

acceleration of ATP resynthesizing sequences until a new equilibrium is reached. More than the absolute level of ATP, the relative contribution of the adenine nucleotides depicted in the adenylate energy charge, AEC = (ATP + $\frac{1}{2}$ ADP)/(ATP + ADP + AMP), is probably of prime importance in the regulation of supply and demand of metabolic energy.[7,8] When this balance is disturbed, ATP is catabolized to hypoxanthine, which enters the extracellular medium from which it can hardly be retaken due to a poor salvage pathway.[9]

The ATP and ADP participating in glycolysis, glycogenolysis, and oxidative phosphorylation are generally termed metabolic ATP and ADP, in contrast to the so-called storage ATP and ADP located in one type of secretion granules, the dense granules.[10-13] Since the dense granule membrane is almost completely impermeable to these compounds or their precursors adenine and adenosine, the storage pool does not take part in energy metabolism, but instead furnishes the ADP which is so important for enhancing aggregation and secretion. In addition, platelets contain a significant amount of actin-bound ADP which can be radiolabeled by ^{14}C-adenine or ^{14}C-adenosine. Although this reflects exchange with metabolic ATP, this pool serves as a separate entity in many different metabolic conditions.[11] Hence, for energetic studies only the metabolic pool of ATP and ADP (about 4.5 and 0.6 μmol · 10^{-11} platelets, respectively[12]) has to be considered, which represents an amount of energy of about 9.5 μmol ATPeq · 10^{-11} platelets. Although other pools of energy are present, most notably glycogen (about 60 μmol glucose residues per 10^{11} cells, which represent 180 μmol ATPeq per 10^{11} cells[14]), rapid exchange with metabolic ATP-ADP does not occur. An exception is the pool of glycolytic intermediates, which represents about 1.3 μmol ATPeq · 10^{-11} platelets. When platelets are treated with H_2O_2, there is a shift of energy-rich phosphates from metabolic ATP-ADP to the sugar phosphates. The glycolytic intermediates then rapidly transfer energy back to metabolic ATP-ADP when a sudden decrease in energy supply threatens to lower these compounds below critical values.[15]

C. Energy Consuming Processes

Equilibrium between supply and demand of energy is essential to preserve cell integrity and to keep the platelets responsive to stimuli. Table 1 shows a number of energy-dependent processes in platelets. The maintenance of secretion granules in particular requires continuous energy support. Platelets have three types of secretion granules: (1) the dense granules containing storage ATP-ADP, Ca^{2+}-ions, and serotonin, (2) the α-granules containing many different proteins, among them coagulation factors and the platelet-specific proteins β-thromboglobulin and platelet factor 4, and (3) the lysosomal or acid hydrolases-containing granules with a variety of lysosomal enzymes, which in compartmentation studies and in certain congenital platelet abnormalities appear to consist of two subtypes (Table 2). Each of these granules has an acidic pH indicating the presence of specific pumps that are driven by ATP hydrolysis[16-18] (a detailed discussion on platelet-dense granules is given in Volume I, Chapter 5). Partial inhibition of ATP hydrolysis for instance by glucose-depletion gradually affects dense granule serotonin, which leaks out of the cell but is rapidly reaccumulated once energy generation is restored. The gradual loss of dense granule ATP-ADP during prolonged storage of platelets for transfusion is also thought to result from a decrease in ATP resynthesizing sequences. A major part of the energy is probably consumed in actin-treadmilling which in resting platelets may use as much as 30 to 50% of the total energy consumption.[19]

III. ENERGY METABOLISM IN STIMULATED PLATELETS

A. ATP-Resynthesizing Pathways

When platelets are stimulated with a high dose of thrombin the cells change shape, aggregate, and undergo complete secretion of dense and α-granules and liberate about 50%

Table 1
ENERGY CONSUMING PROCESSES IN PLATELETS[5]

Type	Unstimulated cells	Stimulated cells
Membrane metabolism		
De novo synthesis of phospholipids		+
Phospholipid methylation	+	−
(Poly)PI turnover	+	+ +
Arachidonate mobilization		+
Receptor processing		+
Endocytosis		
Ion homeostasis		
Na$^+$-K$^+$ ATPase		
Ca^{2+}-Mg^{2+} ATPase	+	−
Na$^+$-H$^+$ antiport		+
Maintenance of membrane potential of		
Plasma membrane	+	−
Dense granule membrane	+	
Lysosomal granule membrane	+	
Transport mechanisms		
Ca^{2+}-uptake	+	+ +
Serotonin accumulation	+	
Adenine/adenosine uptake	+	+ +
Regulatory mechanisms		
cAMP turnover	+	−
cGMP turnover	+	+ +
Protein phosphorylation		
cAMP-dependent kinase	+	−
Ca^{2+}-calmodulin-dependent kinase	−	+
Ca^{2+}-phospholipid-dependent kinase	−	+
Cytoskeleton and cell motility		
Microtubule treadmill		
Actin treadmill	+	+ +
Actin binding protein phosphorylation		+
Myosin phosphorylation	−	+
Actomyosin ATPase	−	+

Symbols used: + = active; + + = accelerated; − = inhibited or inactive.

Table 2
SECRETION GRANULES IN PLATELETS

Secretion Granules

Dense Granules	Storage ATP-ADP, serotonin, PPi, Pi, Ca^{2+}, di-adenosinetetraphosphate
α-Granules	Proteins, e.g., platelet-specific proteins, coagulation factors etc.
Lysosomal granules	Lysosomal enzymes, e.g., β-*N*-acetyl-D-glucosaminidase

α-Granules

Homologs of plasma proteins	Fibrinogen, fibronectin, albumin, Factor V, plasminogen, HMW-kininogen, von Willebrand Factor, CI-inhibitor, factor D + β1H globulin of the complement system, PA-inhibitor, α1-antitrypsin, α2-macroglobulin, α2-antiplasmin, anti-thrombin III, protein S, immunoglobulin G
Platelet-specific	Platelet basic protein, β-thromboglobulin, platelet-derived growth factor, platelet factor 4
Others	Thrombospondin, vascular permeability factor, bactericidal factor, chemotactic factor, CFU-M-inhibitor

Abbreviations used: PA, plasminogen activator; CFU-M, colony forming unit for megakaryocytes.

of the acid hydrolase granule content. These processes are accompanied by a fall in metabolic ATP and AEC, initiation of glycogenolysis, and an increase in glucose uptake, lactate generation, and mitochondrial O_2 uptake.[5,20,21] The onset of the various events is best studied with a combination of optical aggregometry and different types of electrodes, permitting a direct comparison between functional and metabolic changes in a single platelet suspension. In the so-called platelet function analyzer,[6] isolated platelets suspended in a poorly buffered artificial medium are monitored continuously for shape change and aggregation (with a lamp-phototransitor pair), lactate generation (resulting in acidification, monitored with a pH combination electrode), and O_2 consumption (with a micro-Clark electrode). Although absolute data depend on type and concentration of agonist, the general sequence of events following stimulation with 0.1 U $\cdot$ mℓ^{-1} of thrombin is (1) shape change (after 20 sec), (2) an increase in lactate-related proton liberation (after 40 sec), (3) onset of optical aggregation (after 50 sec), and (4) onset of dense-granule secretion (after 70 sec) and thereafter (at higher doses of thrombin) a burst in mitochondrial O_2 uptake. This implies that anaerobic carbohydrate catabolism accelerates before the extracellular appearance of secretion markers, whereas the increase in aerobic energy generation starts later. The increase in glycolytic flux following strong stimulation is about twofold to about 8 μmol lactate $\cdot$ min^{-1} $\cdot$ 10^{-11} cells. During the first 30 sec the burst in mitochondrial respiration reflects a resynthesis of about 5 μmol ATP $\cdot$ min^{-1} $\cdot$ 10^{-11} cells. Hence, a total production of 13 μmol ATP$_{eq}$ $\cdot$ min^{-1} $\cdot$ 10^{-11} cells accompanies the first seconds of platelet activation. Mitochondrial respiration normalizes shortly thereafter but the glycolytic flux remains high until long after platelet functions have been completed. Whether this reflects a de-regulation of flux controlling mechanisms, the so-called disturbance phenomenon remains uncertain. Equally difficult to explain is the independent regulation of anaerobic and aerobic energy support, which is especially evident with weak platelet activating agents that accelerate the glycolytic flux but hardly change mitochondrial respiration.[5] The acceleration of glyco(geno)lysis is due to a change in flux-controlling mechanisms that affect the various enzymes in these sequences, as depicted by rapid changes in glucose-1-phosphate (a product of glycogen phosphorylase), fructose 1,6-bisphosphate (6-phosphofructokinase), and dihydroxyacetone phosphate (aldolase).[22-24] Likely triggers for these changes are the fall in metabolic ATP and AEC, which can be detected as early as 2 sec after platelet stimulation.[20,25] A similar fall in ATP is seen in many other secretory cells, such as neutrophils[26] and mast cells,[27] leading to an increase in ATP resynthesizing processes.[28,29] Of specific interest is the rapid activation of glycogenolytic ATP resynthesis. Activation of glycogen phosphorylase is mediated via a Ca^{2+}-dependent phosphorylase b kinase. An increase in cytosolic Ca^{2+} content is a key event in the intracellular mechanisms that couple receptor occupancy to exocytosis and a link with energy-generating sequences provides an efficient coordination between secretion and regeneration of ATP.

B. Regulation of Supply and Demand of ATP

Induction of secretion is accompanied with a fall in ATP and in the AEC. Although both parameters are closely interrelated, a specific decrease in metabolic ATP that affects the AEC only temporarily can be induced with fluoride, H_2O_2, or transient glucose depletion followed by glucose addition.[7,8,30] Secretion responses then follow the changes in AEC but not those in metabolic ATP indicating that the energy charge plays a major role in the supply of metabolic energy for secretion. The fall in metabolic ATP is steeper with strong agonists (which trigger rapid and maximal secretion), than with weak activators (which induce slow and incomplete responses), indicating that the imbalance between supply and demand is greater as more secretion takes place.[31-34] This suggests that secretion and energy metabolism are tightly connected. Studies on permeabilized cells[35] confirm such a role for energy. Here too, a hydrolyzable form of MgATP is essential to drive dense and lysosomal granule secretion (Figure 2).

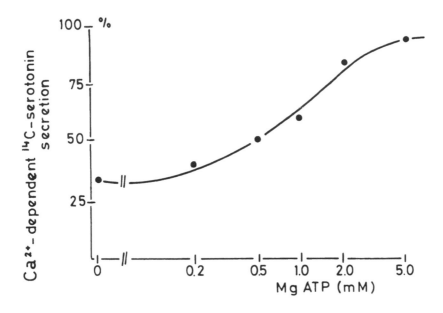

FIGURE 2. MgATP requirement for serotonin secretion. Platelets were subjected to 5 exposures of 20 KV · cm⁻¹ (T = 30 μsec) and 1 min later aliquots were incubated with various concentrations of MgATP. After a further 10 min the cells were challenged with 10 m*M* CaEGTA corresponding to 10⁻⁵ *M* Ca²⁺ and the Ca²⁺-dependent secretion over the next 8 min was determined. (From Knight, D. E. and Scratton, M. C., *Thrombosis Res.*, 20, 437, 1980. With permission.)

It lies within reason that secretion by platelets is sensitive to inhibitors of glycolysis, glycogenolysis, electron transport, and oxidative phosphorylation,[5,22,36-38] as has been observed with adrenal medulla,[39] pancreatic exocrine cells,[40] and growth hormone releasing cells.[41] Since platelets possess Pasteur and Crabtree effects both aerobic and anaerobic ATP resynthesis must be blocked before the metabolic ATP level and AEC decrease and secretion becomes inhibited.[42] This is illustrated by platelets from a patient with a 75% deficiency in hexokinase.[43] Those cells secrete normally as long as mitochondrial energy generation is left intact and can compensate for the slight impairment in glycolytic flux. In the presence of the electron transport inhibitor antimycin A, however, the defect in glycolysis becomes rate limiting and dense granule secretion is retarded. This defect is further enhanced when also glycogenolytic energy production is prevented (Figure 3). A mixture of inhibitors of glyco(geno)lytic and mitochondrial ATP resynthesis blocks secretion by normal platelets after 1 to 2 min, which is apparently the time that is required to exhaust the cell's energy content. When the incubation period is shortened to a few seconds, the decrease in metabolic ATP-ADP is incomplete and the cells remain capable of dense granule secretion.[44] This secretion is then proportional to the remaining metabolic ATP concentration indicating a direct dependence on ATP. A gradual decrease in energy generation can be obtained with suboptimal concentrations of inhibitors, which offers another means to compare secretion with different degrees of energy availability. Those studies reveal that the three types of secretion have different sensitivities to ATP shortage with lysosomal granule secretion showing the greatest sensitivity to impaired energy availability, followed by α-granule secretion, while dense granule secretion is relatively insensitive to a lowering in ATP.[45,46] Hence, the three types of secretion apparently have different energy requirements (Figure 4).

C. Energy Consuming Processes
Table 1 illustrates that a great number of energy consuming processes present in the resting

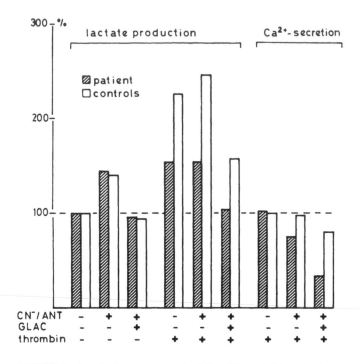

FIGURE 3. Impaired energy generation in hexokinase-deficient platelets. When platelets with a 75% deficiency in hexokinase activity (closed bars) are incubated at 37°C in a glucose and O_2 containing medium, the cells have a normal rate of lactate formation. Addition of cyanide or antimycin A inhibits mitochondrial ATP resynthesis and triggers an increase in the flux from glycogen to lactate. Addition of glucono-δ-lactone inhibits glycogenolysis and glycolysis remains the sole producer of metabolic ATP. Under these conditions the patient's platelets behave normally indicating that 25% remaining hexokinase activity enables normal glycolysis even under stress conditions. When secretion is initiated, however, the energy requirement increases, which is met in part by an acceleration of glycolytic and glycogenolytic flux. In the absence of inhibitors the hexokinase-deficient platelets enhance their glycolytic flux which together with glycogenolytic and mitochondrial energy support enable normal secretion of Ca^{2+} ions from dense granules. In the presence of inhibitors the increase in glycolytic flux is too small and secretion is impaired.

cell show increased activity upon stimulation. Apart from an acceleration in phospholipid metabolism, uptake mechanisms, and phosphorylation of certain proteins, there is an increase in actin treadmilling and actomyosin ATPase activity suggesting that contractile activity becomes a major mechanism for energy utilization.

IV. ENERGY REQUIREMENT OF SECRETION

A. Rapid Blockade of ATP Resynthesis

A better insight in the energy requirement of secretion responses was gained when it became possible to measure energy consumption in intact cells in extremely short-time intervals.[47] The detailed knowledge about energy generating sequences in platelets paved the way for methods that blocked ATP resynthesis in a matter of seconds. Thus, instead of inhibiting platelets before inducing secretion, the cells could be left undisturbed during interaction with a secretagogue and thereafter treated with inhibitors that blocked ATP resynthesis. For such a rapid blockade, two rather contradictory conditions had to be met.

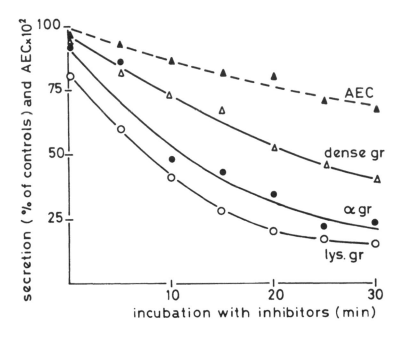

FIGURE 4.   Differential energy requirements for secretion responses by platelets. Platelets were incubated with suboptimal concentrations of antimycin A and 2-deoxy-D-glucose. At various times thereafter, thrombin-induced secretion was determined by analysis of serotonin (dense granules), platelet factor factor 4 (α-granules), and acid hydrolases (lysosomal granules).[46]

First, the cells should maintain ATP homeostasis in order to be optimally responsive to a secretagogue. Second, the cells should be maximally responsive to metabolic inhibitors. In practice, only two approaches led to satisfactory results. In the first approach,[47] platelets were incubated in a medium containing 1 mM CN$^-$ (to block mitochondrial ATP resynthesis) and 1 mM glucose, which just satisfies glucose consumption in conditions where an increased glycolytic flux (with a small contribution of glycogenolysis) compensates for the lack of mitochondrial ATP resynthesis. Then, a mixture of a 30-fold excess of 2-deoxy-D-glucose (an inhibitor of glycolysis) and 10 mM glucono-δ-lactone (an inhibitor of glycogenolysis, final concentrations) blocked glyco(geno)lytic lactate production in less than 5 sec (Figure 5). In the second approach[48] platelets were kept in a glucose-free (and CN$^-$-free) medium, which prevents glycolysis since platelets contain little glucose. ATP homeostasis is then maintained by aerobic degradation of glycogen and, possibly, lipids. Then, a mixture of antimycin A and glucono-δ-lactone blocked mitochondrial O$_2$ uptake and lactate formation almost instantaneously. Both techniques are to some extent supplementary: the CN$^-$ pretreatment blocks mitochondrial ATP resynthesis enabling ATP homeostasis by increased glycolytic flux (anaerobic energy generation) while the pretreatment in glucose-free medium blocks glycolysis enabling ATP homeostasis mainly via increased mitochondrial activity (mainly aerobic energy generation). Both treatments induced a rapid fall in metabolic ATP and, later, in metabolic ADP. When the energy liberated in this decrease in ATP + ADP was expressed as ATP equivalents, a comparison could be made with the energy consumption of undisturbed cells, which could be calculated from the rates of glyco(geno)lysis and oxidative phosphorylation with proper corrections for any change in metabolic ATP and ADP. A 1:1 stoichiometry was found between the energy consumption in platelets with intact ATP resynthesis and cells with blocked ATP regeneration.[47] This relationship was maintained when the temperature was lowered to 10°C or increased to 42°C, indicating that the fall in metabolic ATP-ADP after addition of the inhibitor mixture reflected the energy

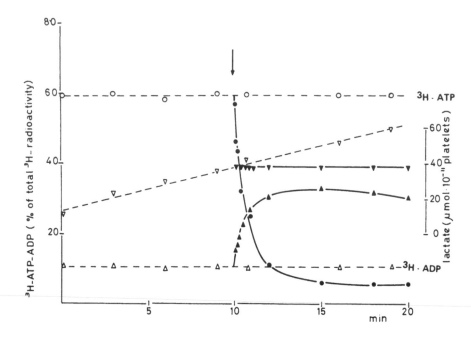

FIGURE 5. Rapid abolishment of ATP resynthesis. When platelets are incubated in a medium with glucose and cyanide, the cells maintain ATP homeostasis by increased glycolytic energy production (open symbols). Addition of a mixture of deoxyglucose, an inhibitor of glycolysis, and glucono-δ-lactone, an inhibitor of glycogenolysis (which starts when glycolysis is inhibited) prevents ATP resynthesis within 5 sec and lactate formation ceases (▼—▼). The ongoing energy consumption leads to rapid hydrolysis of metabolic ATP (●—●), later followed by hydrolysis of metabolic ADP (▲—▲). From these changes the rate of energy consumption can be deduced.

consumption before inducing metabolic blockade. Apparently, during a short period after blocking ATP resynthesis energy consuming processes continue at the same rate as before.[47] The fall in energy content following metabolic arrest therefore provides a sensitive means to quantitate the rate of ATP hydrolysis (Figure 6). With this technique an energy consumption of 3 to 6 μmol ATPeq · min^{-1} · 10^{-11} platelets can be measured in unstimulated cells, which implies that at 37°C the metabolic ATP pool turns over once every minute. The consumption is slightly greater under aerobic conditions (6.2 ± 0.9) than under anaerobic conditions (3.5 ± 0.5, same units) indicating that the loss of mitochondrial energy support leads to a slower energy consumption.

B. ATP Hydrolysis During Secretion

In initial experiments platelets were treated with a combination of metabolic inhibitors and the secretagogue thrombin.[49] Surprisingly, dense granule secretion occurred normally despite the lack of concurrent ATP resynthesis. A closer look revealed that most of the dense granule secretion took place in the period in which the energy content (metabolic ATP + ADP) fell to about 10% of initial levels. Thus, in the absence of ATP resynthesis the platelets contain enough energy to support secretion.[49] The discovery that the platelet's energy content contained sufficient energy to drive secretion implied that the fall in metabolic ATP-ADP after abolishing ATP resynthesis was in some way related to the secretion process. When thrombin (5 U · mℓ^{-1}) is added, secretion of dense, α- and lysosomal granules follow, and the cells change in shape and aggregate (Figure 7). These functions are accompanied by a two- to threefold increase in energy consumption during the first 15 sec after stimulation.[47] Thereafter, energy consumption decreases returning to values of unactivated plaelets after about 45 sec. Following stimulation with weak agonists or low doses of thrombin the

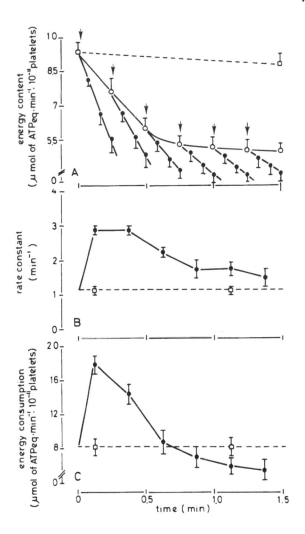

FIGURE 6. Energy parameters in platelets with blocked ATP resynthesis. Platelets, incubated in glucose-free medium maintain ATP-homeostasis, as depicted by a stable energy content (Figure 6A, □—□). The secretagogue thrombin (5 U/mℓ, final concentration) disturbs the balance between supply and demand and induces a fall in energy content until a new equilibrium is reached. (○—○) Abolishment of ATP regeneration by a mixture of antimycin A-glucono-δ-lactone to the stimulated cells (arrows) reveals the rate of energy consumption at that stage (●—●). The fall in energy content follows a single exponential model from which the rate constant can be derived (Figure 6B). Figure 6C illustrates the rate of energy consumption at various intervals of unstimulated and thrombin-stimulated platelets. The thrombin-induced changes in energy consumption reflect both the changes in metabolic ATP-ADP content (illustrated by the broken line in Figure 6A) and changes in the rate of ATP-ADP turnover (illustrated in Figure 6B). (From Verhoeven, A. J. M. et al., *Int. J. Biochem.*, 18, 985, 1986. With permission.)

energy consumption normalizes in the range of resting platelets. With strong stimulation, however, the energy consumption stabilizes in a range below the values found in resting cells. Hence, the rate of ATP resynthesizing pathways in platelets that have completed secretion depends on the rate with which the secretion responses have been carried out.

On first sight there appears to be a close correlation between the rate of secretion and the rate of concurrent energy consumption: when secretion slows down, less energy is consumed. A similar correlation is also found when low doses of thrombin are used leading to slower

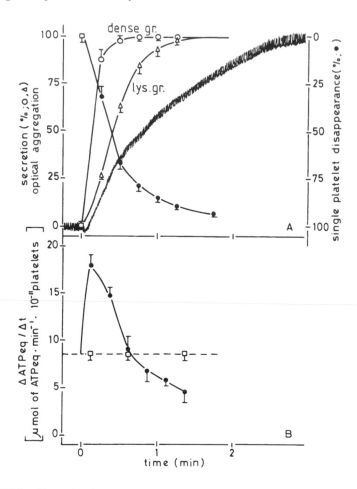

FIGURE 7.   Comparison between aggregation and secretion and the concurrent consumption of metabolic energy. Platelets were stimulated with thrombin (5.0 U · mℓ⁻¹ final concentration) and optical aggregation was monitored. At various times samples were collected for analysis of single platelet disappearance (●) and secretion (expressed as percentage of maximal secretable amount of marker) of ¹⁴C-serotonin (○) and N-acetyl-β-D-glucosaminidase (Δ)(A). Also shown are the values for secretion and single platelet disappearance at 5 sec before addition of thrombin (□). Data on secretion of β-thromboglobulin were always intermediate between those of ¹⁴C-serotonin and the acid hydrolase; these data were omitted for the sake of clarity. Concurrently (B) energy consumption, expressed as μmol of ATPeq · min⁻¹ · 10⁻¹¹ platelets was measured both in thrombin stimulated (●—●) and unstimulated (□—□) suspensions. Data are expressed as means ± SD (n = 4). (From Verhoeven, A. J. M. et al., *Biochem. J.*, 221, 777, 1984. With permission.)

secretion and proportionally slower energy consumption. Similarly, when the initial level of metabolic ATP is decreased by pretreatment with metabolic inhibitors or when the temperature is lowered, slower secretion and slower energy consumption follow. Thus, under a variety of different metabolic and functional conditions the coupling between secretion velocity and energy consumption is maintained. The energy made available during the hydrolysis of ATP is represented by Equation 1:

$$\Delta G_p = \Delta G_p^{\circ\prime} + RT \cdot \ln ([Pi][ADP]/[ATP]) \qquad (1)$$

It follows from Equation 1 that changes in ATP level and temperature affect the amount of energy liberated during ATP hydrolysis, which may alter the energy vs. secretion relation-

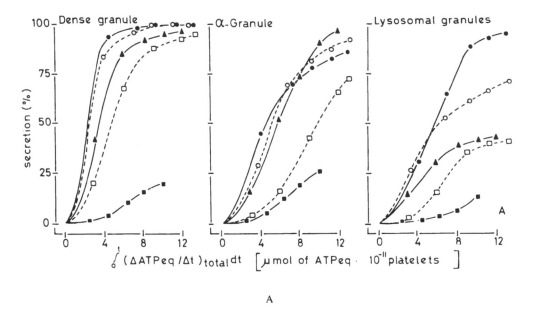

A

FIGURE 8. Comparison between energy consumption and extent of secretion. The amount of energy, expressed as μmol of ATPeq $\cdot$ 10^{-11} platelets, that is consumed from the moment of stimulation was calculated as: $\int_{0}^{t}$ $(\Delta ATPeq/\Delta t)_{total}$ dt from the fall of metabolic ATP/ADP after metabolic arrest in 15-sec intervals and plotted vs. the corresponding extent of secretion of ^{14}C-serotonin (dense granule), β-thromboglobulin (α-granule), and N-acetyl-β-D-glucosaminidase (lysosomal granules). The symbols are 0.05 (■—■), 0.1 (□---□), 0.2 (▲—▲), 0.5 (○---○), and 5.0 (●—●) units of thrombin $\cdot$ mℓ^{-1}. (A) Total energy consumption in thrombin-treated platelets; (B) Incremental energy consumption (difference between thrombin-stimulated and unstimulated platelets). (From Verhoeven, A. J. M. et al., *Biochem. J.*, 221, 777, 1984. With permission.)

ships. However, no such effect shows up and the quantitative correlation between energy consumption and secretion velocity appears to hold within a wide range of initial ATP levels, temperatures, and concentrations of different types of platelet agonists.[47,48,50]

C. Separation Between Total and Incremental Energy Consumption

A major uncertainty in the comparison between secretion and concurrent energy consumption is whether only the extra energy consumption seen during secretion should be considered or the total energy consumption should be taken into account. A comparison between secretion of dense, α-, and lysosomal granules and either total energy consumption (Figure 8A) or incremental energy consumption (Figure 8B) reveals a much better correlation with the latter. This suggests that the energy consumption present before addition of the secretagogue, the so-called basal energy consumption, continues largely unchanged in the period in which secretion is carried out. This is an important observation since especially in the early part of secretion the total energy consumption is for the major part determined by the basal energy consumption since the increment in energy consumption at this stage is relatively small (Figure 8). Furthermore, it suggests that a major part of the energy consuming processes that are present before stimulation continue unaltered forming a metabolic entity that is independent of the processes involved in secretion. An exact estimation of the relative roles of basic and incremental energy consumption is difficult to make when strong secretagogues are used. Here, energy consumption stabilizes after secretion at levels below the range found before stimulation indicating that the basal energy consumption decreases during the secretion process. Apparently, in this case the increment in energy consumption insufficiently satisfies the energy need for secretion and basal and new ATP-consuming processes

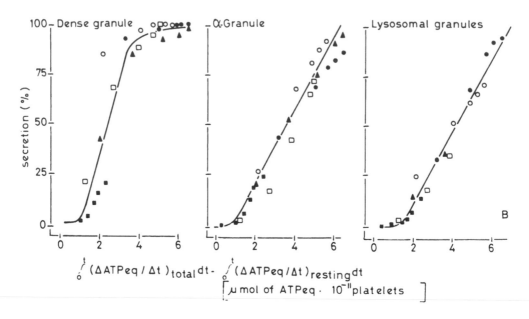

FIGURE 8B.

have to compete for available ATP. Possibly, there is a gradual decrease in basic consumption during secretion but an abrupt fall immediately after stimulation cannot be excluded.

D. Secretion Induced by Different Secretagogues

The secretion vs. energy relationship established with thrombin is also found when collagen is used as a secretagogue.[51] Other stimulators such as the bivalent cathionophore A23187 or the weak activator ADP lead to different correlations. A23187-induced acid hydrolase secretion costs about six times more energy than thrombin-induced secretion, apparently because the ionophore affects a number of ATPases that are irrelevant for secretion but increases the rate of ATP hydrolysis. ADP induces a biphasic pattern in the rate of energy consumption clearly illustrating the separate energy requirements of aggregation and secretion (Figure 9). The energy consumed during the secretion phase is about twofold higher than that found with thrombin or collagen, a difference which at present remains unexplained.

The comparisons made so far between secretion and energy consumption all point to the concept that a certain amount of secretion correlates with a specific amount of energy consumption. The need for energy appears to be the same with different types of agonists but depends on the rate with which secretion takes place. These observations have led to the concept of the energy quotient for secretion, which is defined as the metabolic price for the extrusion of 1% of the granule contents.[47,51] With this concept the metabolic price for a separate secretion response can be calculated despite the fact that secretion of dense, α-, and acid hydrolase granules greatly overlaps. The basis of these calculations is the assumption that the coupling between a single type of secretion and energy is constant at any stage of the response whether or not other secretion responses occur in parallel. The energy quotient for acid hydrolase secretion can be measured in a period where dense- and α-granule secretion have been completed. Assuming that this quotient is similar in a stage where acid hydrolase secretion is accompanied by α-granule secretion, the energy quotient measured in this stage can be corrected for the contribution of acid hydrolase secretion, the difference being the quotient for the separate secretion of α-granules. Similarly, the quotient for dense granule secretion can be calculated. Extrapolation to 100% secretion then results in the metabolic price for a separate secretion response (Table 3). The cost for complete secretion of dense

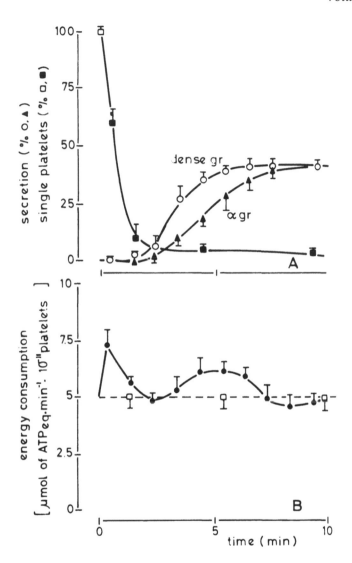

FIGURE 9. Comparison between aggregation and secretion and concurrent energy consumption in ADP-stimulated platelets. ^{3}H-adenine and ^{14}C-serotonin-labeled platelets were stimulated with 100 μM ADP (final concentration) and samples were collected at the indicated times for analysis of functional responses (A) and concurrent energy consumption (B). In (A), the data of single platelet disappearance (■) and secretion of ^{14}C-serotonin (○) and β-thromboglubulin (▲) are shown. Concurrently, energy consumption, expressed as μmol of ATPeq $\cdot$ min^{-1} $\cdot$ 10^{-11} platelets was measured in both stimulated (●—●) and unstimulated (□---□) suspensions. Data are expressed as means ± SD (n = 3). (From Verhoeven, A. J. M. et al., *Biochem. J.*, 236, 879, 1986. With permission.)

granules is small ranging narrowly between 0.5 and 0.8 μmol ATPeq $\cdot$ 10^{-11} platelets with all agonists tested. Much higher is the price for the combined secretion of α- and acid hydrolase granules (which so far can be insufficiently separated) leading to a cost of about 6 μmol ATPeq $\cdot$ 10^{-11} platelets with thrombin and collagen but almost twice as much with ADP. Much higher data are found with A23187 but in view of the many side effects of this compound this number is probably an overestimation. The great similarity in energy data obtained with different agonists is of special interest since each agonist is known to activate

Table 3
THE METABOLIC PRICE FOR SECRETION RESPONSES BY PLATELETS

Agonist	Energy quotient		Energy cost for complete secretion	
	Dense granules	α-Granules + lysosomal granules	Dense granules	α-Granules + acid hydrolase granules
Thrombin	0.007	0.053	0.7	5.3
Collagen	0.008	0.069	0.8	6.9
ADP	0.005	0.112	0.5	11.2
A23187	0.006	0.323	0.6	32.3

Table 4
EFFECT OF INHIBITORS[51]

Addition	Increase in energy consumption (%)	Dense granule secretion (%)
None	100	100
dBcAMP	46 ± 13	14 ± 1
Theophylline-PGE₁	34 ± 14	12 ± 1
Cytochalasin B	112 ± 17	106 ± 7
Colchicine	112 ± 17	67 ± 4
Indomethacin	91 ± 12	99 ± 2
Ouabain	103 ± 13	95 ± 6
Quercetin	100 ± 17	93 ± 8

the platelets via different mechanisms. Apparently the major part of the energy cost is primarily determined by processes common to all agonists which are probably at a final stage in stimulus-secretion coupling.

E. Cellular Energy Metabolism vs. Separate ATP-Dependent Reactions

At present it is uncertain where the energy that is so essential for secretion is used for. The exact mechanisms that couple receptor binding of a secretagogue with the extrusion of granule contents are far from understood. Nevertheless, a number of energy-requiring processes have been identified such as receptor capping,[52,53] transbilayer movement of phospholipids,[54] phosphatidylinositol cycle and metabolism of polyphosphoinositides,[55-57] Ca^{2+} sequestration,[58-61] protein kinase C activity,[62,63] and most importantly actomyosin contractile activity.[64,65] In an attempt[51] to identify the major energy consuming processes in this sequence a number of inhibitors have been evaluated on their effect on dense granule secretion and concurrent energy consumption induced by a high dose of thrombin (Table 4). Agents that raise cAMP such as dibutyryl-cAMP and a combination of theophylline and PGE₁ greatly abolish secretion as well as a major part of the increment in energy consumption. The cytoskeleton disruption agents cytochalasin B and colchicine only inhibit secretion in the case of colchicine without changing the increase in energy consumption. Inhibitors of prostaglandin endoperoxide-thromboxane A_2 formation (indomethacin) or Na^+-K^+ ATPase activity (ouabain, quercetin) have no effect. The influence of the inhibitors is greater when a low dose of thrombin is used, but again a reduction in secretion is paralleled by a reduction in energy consumption. The only exception is colchicine which inhibits secretion without affecting energy consumption. Thus, the strongest effect is seen with agents that raise cAMP but since many different processes are susceptible to changes in cAMP content a separate

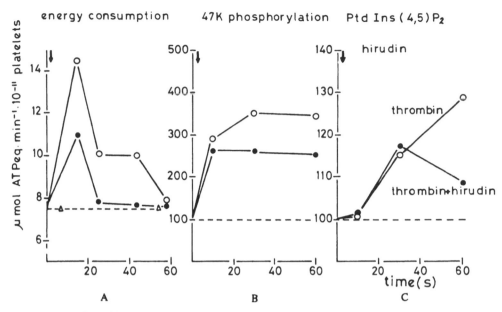

FIGURE 10. Effect of hirudin on thrombin-induced alterations in energy consumption, 47,000-dalton phospho-rylation, and PtdIns(4,5)P$_2$ phosphorylation. ^{32}P P$_i$-labeled gel-filtered platelets were incubated either with thrombin (0.2 U · mℓ$^{-1}$) alone (○), or with 10 U · mℓ$^{-1}$ of hirudin added at 2 sec after thrombin (arrows); ●). At the times indicated, samples were collected for the analysis of energy consumption (μmol ATPeq · min^{-1} · 10^{-11} cells), ^{32}P-phosphorylation of the 47-kdalton protein, and ^{32}P-labeled phosphatidylinositol bisphosphate, PtdIns (4,5)P$_2$ (% of controls).

energy consuming process cannot be identified. Two likely candidates for consumption of a probably minor amount of energy are changes in polyphosphoinositides and the phospho-rylation of two specific proteins, with molecular weights of 47,000 and 20,000, being the substrates of protein kinase C[62] and myosin light chain kinase,[64] respectively. Figure 10A illustrates the increase in energy consumption following stimulation with 0.2 U · mℓ$^{-1}$ thrombin. The increase is accompanied with rapid phosphorylation of the 47 K protein (Figure 10B) and the 20 K protein (not shown) as well as PtdIns(4,5)P$_2$ (Figure 10C), PtdIns(4)P, and PtdA* formation (not shown) and results in 85% dense granule secretion after 1 min. When the stimulation by thrombin is rapidly abolished by adding an excess of hirudin as soon as 2 sec after thrombin addition, energy consumption rapidly normalizes and no secretion takes place. PtdIns(4,5)P$_2$ and PtdIns(4)P return to normal levels but the phosphorylation of 47 K and 20 K proteins remains high. Hence, the thrombin-hirudin treatment leads to a transient increase in energy consumption with concomitant changes in ^{32}P-labeled polyphosphoinositides and proteins suggesting that these events are closely cou-pled. The result after 1 min is that platelets have a normal energy consumption and normal levels of ^{32}P-polyphosphoinosites but still maintain elevated levels of ^{32}P-proteins.[66] One may speculate that such an early investment in energy consuming processes leads to a cheaper response when secretion is initiated by a second stimulation with thrombin. As illustrated in Figure 11 normal dense granule secretion can be induced in cells that have undergone a transient stimulation by the thrombin-hirudin treatment.[66] The relation between secretion and incremental energy consumption is again linear as seen in normal platelets, but, un-expectedly, this relationship has shifted to higher energy data. Hence, in spite of the early investment of energy which is not accompanied by secretion, subsequent secretion requires

* Abbreviations used are phosphatidylinositol (PtdIns), phosphatidylinositol 4-monophosphate (PtdIns(4)P), phos-phatidylinositol 4,5-biphosphate (PtdIns)4,5)P$_2$), and phosphatidic acid (PtdA).

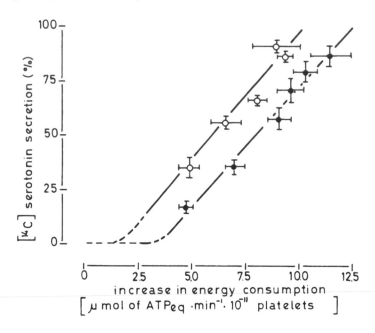

FIGURE 11. Relationship between energy consumption and dense granule secretion induced by various doses of thrombin with and without prior transient stimulation. ^{3}H-adenine- and ^{14}C-serotonin-labeled gel-filtered platelets were stimulated with thrombin (0.2 U · mℓ$^{-1}$). At 2 sec thereafter, thrombin was neutralized by the addition of 5 U · mℓ$^{-1}$ of hirudin. At 60 sec the platelets were stimulated with a second dose of thrombin (final concentrations ranging between 4.8 and 10 U · mℓ$^{-1}$). Secretion of ^{14}C-serotonin at 15 sec after the second dose (expressed as a percentage of maximal secretable) and concurrent energy consumption during the first 15-sec interval (expressed as μmol of ATPeq · min^{-1} · 10^{-11} platelets, was determined (●). The control suspensions consisted of platelets, which were incubated for 60 sec with hirudin (5 U · mℓ$^{-1}$) without prior activation; then different doses of thrombin were added to final concentrations of 4.8 to 10 U · mℓ$^{-1}$ (○). Energy consumption was corrected for (ΔATPeq/Δt) measured simultaneously in unstimulated suspensions (7.4 ± 0.9 μmol of ATPeq · min^{-1} · (10^{11} platelets)$^{-1}$; n = 8). Data are means ± SD from 24 incubations. The slopes of both lines differed not significantly $p > 0.25$; the difference between both intercepts was highly significant ($p < 0.005$), as determined by an F-distribution-test. (From Verhoeven, A. J. M. et al., *Biochem. J.*, 228, 451, 1985. With permission.)

more energy than during a normal secretion response. The difference is caused by an increase in the threshold below which energy consumption is increased without concomitant secretion. Apparently, prestimulation below a certain threshold activates mechanisms that make the cell less responsive to a second stimulation resulting in a higher energy need for execution of secretion.

V. CONCLUSION

The fall in energy content immediately following abrupt and complete abolishment of ATP-resynthesis reflects the energy consumption of undisturbed platelets. Unstimulated platelets consume 3 to 6 μmol ATPeq · min^{-1} · 10^{-11} cells (37°C) depending on whether the mitochondria contribute to ATP regeneration or not. For a cell that lacks considerable *de novo* synthesis of proteins, lipids, or complex carbohydrates this number is high and several-fold more than in other blood cells. This has led to the hypothesis that before stimulation platelets build up a storage of energy that can be rapidly released once the cells are activated ("spring theory", Volume I, Chapter 6). Subsequent platelet responses such as aggregation and secretion would then occur independently from concurrent ATP resynthesizing pathways. Present evidence, however, argues against such a concept. There is a strong coupling between the rate of secretion responses and concurrent ATP hydrolysis in platelets stimulated with different agonists, at different temperatures, and at different initial ATP concentrations. This indicates that ATP hydrolysis is directly used for support of the different secretion responses. The coupling is particularly evident when only the increment in energy consumption is considered, suggesting that the major part of the basal consumption is less important for secretion.

The energy cost of dense granule secretion is about tenfold lower than the cost of the combined secretion of α- and lysosomal granules. Gradual inhibition of ATP regeneration inhibits lysosomal granule secretion at an earlier stage than α-granule secretion, whereas dense granule secretion is the least sensitive to metabolic inhibitors.[46] This suggests that the energy cost of α-granule secretion is intermediate between the prizes of lysosomal and dense granule secretion. The fact that different types of secretion require different amounts of energy is unexpected. An uncertainty in these comparisons is how secretion must be measured. Ideally, the three types of secretion granules should be (1) equal in number, (2) equal in size, and (3) contain a marker that is specific for one type of granule but not different from the markers in other granules in terms of chemical and physical properties. The real situation is quite different. Dense granules are small (250 to 300 nm in diameter) and few in number (about six per platelet), each containing a limited amount of low molecular weight substances such as serotonin, nonmetabolic ATP-ADP, and Ca^{2+}-ions.[67] In contrast, a single platelet has many α-granules, each 300 to 500 nm in diameter, which contain a great variety of proteins such as platelet-specific proteins, coagulation factors, and von Willebrand Factor, a cofactor for platelet adhesion to subendothelium with a molecular weight of more than 1 million daltons.[68,69] The number of lysosomal granules is uncertain but here too a great number of proteins are stored. There is not much difference between the secretion characteristics among dense granule constituents, as measured by chemiluminescence (ATP-secretion), Ca^{2+}-indicators, or a Ca^{2+}-electrode (Ca^{2+} secretion) and endogenous or exogenous serotonin.[70-72] In contrast, there is strong heterogeneity in the secretion properties of α-granule markers, varying from slight discrepancies between β-thromboglobulin and platelet factor 4[45,73] to completely different secretion patterns such as with fibronectin[74] and von Willebrand Factor[75] which are only slowly extruded. A complicating factor with acid hydrolase secretion is the fact that even with strong stimulation, secretion is far from complete with retained fractions ranging from 50% (β-N-acetylglucosaminidase) to 70% (β-galactosidase) of total contents. This percentage can be greatly reduced by preincubation with lysosomotropic agents[18] but the energetic consequences of such a treatment have been left unexplored.

ACKNOWLEDGMENTS

The author gratefully acknowledges the cooperation with Dr. A. J. M. Verhoeven, Mrs. G. Gorter, and M. Mommersteeg and the secretarial assistance of Mrs. A. Beyer and M. Hoeneveld.

Part of this work has been supported by The Netherlands Organization for the Advancement of Pure Research (ZWO), FUNGO grants 13-30-36 and 13-30-60, NATO-research grant 1903, and grant 82.055 of the Dutch Heart Foundation.

REFERENCES

1. **Kinlough-Rathbone, R. L., Packham, M. A., and Mustard, J. F.,** The effect of glucose on the platelet response to release-inducing stimuli, *J. Lab. Clin. Med.,* 80, 247, 1972.
2. **Muenzer, J., Weinbuch, E. C., and Wolfe, S. M.,** Oxygen consumption of human blood platelets. Effect of thrombin, *Biochim. Biophys. Acta,* 376, 237, 1975.
3. **Akkerman, J. W. N., Gorter, G. and Sixma, J. J.,** Regulation of glycolytic flux in human platelets: relation between energy production by glyco(geno)lysis and energy consumption, *Biochim. Biophys. Acta,* 541, 241, 1978.
4. **Meltzer, H. Y. and Guschwan, A.,** Type 1 (brain type) creatine phosphokinase (CPK) activity in rat platelets, *Life Sci.,* 11, 121, 1972.
5. **Akkerman, J. W. N. and Verhoeven, A. J. M.,** Energy Metabolism and function, in *Platelet Responses and Metabolism,* Vol. 3, Holmsen, H., Ed., CRC Press, Boca Raton, Fla., 1987, 69.
6. **Akkerman, J. W. N. and Holmsen, H.,** Interrelationships among platelet responses. Studies on the burst in proton liberation, lactate production and oxygen uptake during platelet aggregation and Ca^{2+} secretion, *Blood,* 57, 956, 1981.
7. **Holmsen, H. and Robkin, L.,** Hydrogen peroxide lowers ATP levels in platelets without altering adenylate energy charge and platelet function, *J. Biol. Chem.,* 252, 1752, 1977.
8. **Akkerman, J. W. N., Gorter, G., and Holmsen, H.,** Reversibility of the inhibition of platelet responses by ATP-deprivation. Close correlation between responses and adenylate energy charge during transient substrate depletion, *Biochim. Biophys. Acta,* 760, 34, 1983.
9. **Rivard, G. E., McLaren, J. D., and Brunst, R. F.,** Incorporation of hypoxanthine into adenine and guanine nucleotides by human platelets, *Biochim. Biophys. Acta,* 381, 144, 1975.
10. **Akkerman, J. W. N., Ebberink, R. H. M., Lips, J. P. M., and Christiaens, G. C. M. L.,** Rapid separation of cytosol and particle fraction of human platelets by digitonin-induced cell damage, *Br. J. Haematol.,* 44, 291, 1980.
11. **Daniel, J. L., Molish, I. R., and Holmsen, H.,** Radioactive labeling of the adenine nucleotide pool of cells as a means to distinguish among intracellular compartments. Studies on human platelets, *Biochim. Biophys. Acta,* 632, 444, 1980.
12. **Akkerman, J. W. N., Nieuwenhuis, H. K., Mommersteeg-Leautaud, M. E., Gorter, G., and Sixma, J. J.,** ATP-ADP compartmentation in Storage Pool Deficient platelets. Correlation between granule bound ADP and the bleeding time, *Br. J. Haematol.,* 55, 135, 1983.
13. **Lages, B., Holmsen, H., Weiss, H. J., and Dangelmaier, C.,** Thrombin and ionophore A23187-induced dense granule secretion in storage pool deficient platelets: evidence for impaired nucleotide storage as the primary dense granule defect, *Blood,* 61, 154, 1983.
14. **Akkerman, J. W. N.,** Carbohydrate metabolism, in *Platelet Responses and Metabolism,* Vol. 2, Holmsen, H., Ed., CRC Press, Boca Raton, Fla., 1987, 189.
15. **Verhoeven, A. J. M., Mommersteeg, M. E., and Akkerman, J. W. N.,** Balanced contribution of glycolytic and adenylate pool in supply of metabolic energy in platelets, *J. Biol. Chem.,* 260, 2621, 1985.
16. **Johnson, R. G., Scarpa, A., and Salganicoff, L.,** The internal pH of isolated serotonin containing granules of pig platelets, *J. Biol. Chem.,* 253, 7061, 1978.
17. **Fishkes, H. and Rudnick, G.,** Bioenergetics of serotonin transport by membrane vesicles derived from platelet dense granules, *J. Biol. Chem.,* 257, 5671, 1982.
18. **Van Oost, B., Smith, J. B., Holmsen, H., and Vladutiu, G. D.,** Lysosomotropic agents potentiate thrombin-induced acid hydrolase secretion from platelet, *Proc. Natl. Acad. Sci. U.S.A.,* 82, 2374, 1985.
19. **Daniel, J. L., Molish, I. R., Robkin, L., Holmsen, H.,** Nucleotide exchange between cytosolic ATP and F-actin-bound ADP may be a major energy utilizing process in unstimulated platelets, *Eur. J. Biochem.,* 156, 677, 1986.
20. **Fukami, M. H., Holmsen, H., and Salganicoff, L.,** Adenine nucleotide metabolism of blood platelets. IX. Time course of secretion and changes in energy metabolism in thrombin-treated platelets, *Biochim. Biophys. Acta,* 444, 633, 1976.
21. **Mürer, E. H.,** Release reaction and energy metabolism in blood platelets with special reference to the burst in oxygen uptake, *Biochim. Biophys. Acta,* 162, 320, 1968.
22. **Detwiler, T. C.,** Control of energy metabolism in platelets. The effects of thrombin and cyanide on glycolysis, *Biochim. Biophys. Acta,* 256, 163, 1972.
23. **Akkerman, J. W. N., Gorter, G., Sixma, J. J., and Staal, G. E. J.,** Variations in levels of glycolytic intermediates in human platelets during platelet-collagen interaction, *Biochim. Biophys. Acta,* 421, 296, 1976.
24. **Akkerman, J. W. N., Driver, H. A., Dangelmaier, C. A., and Holmsen, H.,** Alterations in ^{32}P-labeled intermediates during flux activation of human platelet glycolysis, *Biochim. Biophys. Acta,* 805, 221, 1984.
25. **Mills, D. C. B.,** Changes in the adenylate energy charge in human blood platelets induced by adenosine diphosphate, *Nature (London),* 243, 220, 1973.

26. **Borregaard, N. and Herlin, T.**, Energy metabolism in human neutrophils during phagocytosis, *J. Clin. Invest.*, 70, 550, 1982.

27. **Johansen, T.**, Ethacrynic acid inhibition of histamine release from rat mast cells: effect on cellular ATP levels and thiol groups, *Eur. J. Pharmacol.*, 92, 181, 1983.

28. **Rossignol, B., Herman, G., Chambout, A. M., and Keryer, G.**, The calcium ionophore A23187 as a probe for studying the role of Ca²⁺ ions in the mediation of carbochol effects on rat salivary glands: protein secretion and metabolism of phospholipids and glycogen, *FEBS Lett.*, 43, 241, 1968.

29. **Malaisse, W. J., Sener, A., Koser, M., Ravazolla, M., and Malaisse-Lagaue, F.**, The stimulus-secretion coupling of glucose-induced insulin release. Insulin release due to glycogeneolysis in glucose-deprived islets, *Biochem. J.*, 164, 147, 1977.

30. **Mürer, E. H., Day, H. J., and Lieberman, J. E.**, Metabolic aspects of the secretion of stored compounds from blood platelets. III. Effect of NaF on washed platelets, *Biochim. Biophys. Acta*, 362, 266, 1974.

31. **Holmsen, H., Day, H. J., Setkowsky, C. A.**, Secretory mechanisms. Behaviour of adenine nucleotides during the platelet release reaction induced by adenosine diphosphate and adrenaline, *Biochem. J.*, 129, 67, 1972.

32. **Akkerman, J. W. N., van Brederode, W., Gorter, G., Zegers, B. J. M., and Kuis, W.**, The Wiskott-Aldrich syndrome: studies on a possible defect in mitochondrial ATP resynthesis in platelets, *Br. J. Haematol.*, 51, 561, 1982.

33. **Holmsen, H.**, Biochemistry of the platelet: energy metabolism, in *Haemostasis and Thrombosis, Basic Principles and Clinical Practice*, Colman, R. W. et al., Eds., B. Lippincott, Philadelphia, 1982, 431.

34. **Sanchez, A.**, Ca²⁺-independent secretion is dependent on cytoplasmic ATP in human platelets, *FEBS Lett.*, 191, 283, 1985.

35. **Knight, D. E. and Scrutton, M. C.**, Direct evidence for a role for Ca²⁺ in amine storage granule secretion by human platelets, *Thrombosis Res.*, 20, 437, 1980.

36. **Holmsen, H., Setkowsky, C. A., and Day, H. J.**, Effects of antimycin and 2 deoxy-glucose on adenine nucleotides in human platelets, *Biochem. J.*, 144, 385, 1974.

37. **Katlove, H. and Gomez, M. H.**, Collagen-induced platelet aggregation. The role of adenine nucleotides and the release reaction, *Thromb. Diath. Haemorrh.*, 34, 795, 1975.

38. **Verhoeven, A. J. M., Mommersteeg, M. E., and Akkerman, J. W. N.**, Metabolic energy is required in human platelets at any stage during optical aggregation and secretion, *Biochim. Biophys. Acta*, 800, 242, 1984.

39. **Rubin, R. P.**, The role of energy metabolism in calcium evoked secretion from the adrenal medulla, *J. Physiol.*, 206, 181, 1970.

40. **Jamieson, J. D. and Palade, G. E.**, Synthesis, intracellular transport and discharge of secretory proteins in stimulated pancreatic exocrine cells, *J. Cell Biol.*, 50, 135, 1971.

41. **McPherson, M. and Schofield, J. C.**, Requirement for adenosine triphosphate for stimulation in vitro of ox growth-hormone release, *Biochem. J.*, 140, 479, 1974.

42. **Mürer, E. H. and Holme, R.**, A study of the release of calcium from human blood platelets and its inhibition by metabolic inhibitors, N-ethylmaleimide and aspirin, *Biochim. Biophys. Acta*, 22, 197, 1970.

43. **Akkerman, J. W. N., Rijksen, G., Gorter, G., and Staal, G. E. J.**, Platelet functions and energy metabolism in a patient with hexokinase deficiency, *Blood*, 63, 147, 1984.

44. **Holmsen, H. and Akkerman, J. W. N.**, The requirement for ATP availability in platelet response. A quantitative approach, in *The Regulation of Coagulation*, Mann, K. G. and Taylor, F. B., Eds., Elsevier-North Holland, Amsterdam, 1980, 409.

45. **Holmsen, H., Robkin, L., and Day, H. J.**, Effects of antimycin A and 2-deoxyglucose on secretion in human platelets. Differential inhibition of the secretion of acid hydrolases and adenine nucleotides, *Biochem. J.*, 182, 413, 1979.

46. **Holmsen, H., Kaplan, K. L., and Dangelmaier, C. A.**, Differential energy requirements for platelet responses, *Biochem. J.*, 208, 9, 1982.

47. **Akkerman, J. W. N., Gorter, G., Schrama, L. H., and Holmsen, H.**, A novel technique for rapid determination of energy consumption in platelets. Demonstration of different energy consumption associated with three secretory responses, *Biochem. J.*, 210, 145, 1983.

48. **Verhoeven, A. J. M., Mommersteeg, M. E., and Akkerman, J. W. N.**, Quantification of energy consumption in platelets during thrombin-induced aggregation and secretion. Tight coupling between platelets responses and the increment in energy consumption, *Biochem. J.*, 221, 777, 1984.

49. **Akkerman, J. W. N., Holmsen, H., and Driver, H. A.**, Platelet aggregation and Ca²⁺-secretion are independent of simultaneous energy production, *FEBS Lett.*, 100, 286, 1979.

50. **Verhoeven, A. J. M., Mommersteeg, M. E., and Akkerman, J. W. N.**, Kinetics of energy consumption in human platelets with blocked ATP regeneration, *Int. J. Biochem.*, 18, 985, 1986.

51. **Verhoeven, A. J. M., Mommersteeg, M. E., and Akkerman, J. W. N.**, Comparative studies on the energetics of platelet responses induced by different agonists, *Biochem. J.*, 236, 879, 1986.

52. **Bourguignon, L. Y. W.,** Receptor capping in platelet membranes, *Cell Biol. Int. Rep.,* 8, 19, 1984.

53. **Stubbs, C. D. and Smith, A. D.,** The modification of mammalian membrane polyunsaturated fatty acid composition in relation to membrane fluidity and function, *Biochim. Biophys. Acta,* 779, 89, 1984.

54. **Seigneuret, M. and Devaux, P. F.,** ATP-dependent asymmetric distribution of spin-labeled phospholipids in the erythrocyte membrane: relation to shape changes, *Proc. Natl. Acad. Sci. U.S.A.,* 81, 3751, 1981.

55. **Haslam, R. J. and Davidson, M. M. L.,** Potentiation by thrombin of the secretion of serotonin from permeabilized platelets equilibrated with Ca^{2+} buffer, *Biochem. J.,* 222, 351, 1984.

56. **Holmsen, H., Nilsen, A. O., and Rongved, S.,** Energy requirements for stimulus-response coupling, in *Mechanisms of Stimulus-Response Coupling in Platelets,* Westwick, J., Scully, M. F., MacIntyre, D. E., and Kakker, V. V., Eds., Plenum Press, 1985, 215.

57. **Holmsen, H., Dangelmaier, C. A., and Rongved, S.,** Tight coupling of thrombin-induced acid hydrolase secretion and phosphatidate synthesis to receptor occupancy in human platelets, *Biochem. J.,* 222, 157, 1984.

58. **De Metz, M., Enouf, J., Lebret, M., and Lévy-Toledano, S.,** The Ca^{2+}-uptake and the hydrolysis of various nucleotide triphosphates by human platelet membranes, *Biochim. Biophys. Acta,* 773, 325, 1984.

59. **Levitsky, D. O., Loginov, V. A., Lebedev, A. V., Levchenki, T. S., and Leytin, V. L.,** Ca^{2+}-binding and charge movements in membranes of platelets and sarcoplasmic reticulum, *FEBS Lett.,* 171, 89, 1984.

60. **Purdon, A. D., Daniel, J. L., Stewart, G. J., and Holmsen, H.,** Cytoplasmic free calcium concentration in porcine platelets, *Biochim. Biophys. Acta,* 800, 178, 1984.

61. **Menashi, S., Authi, K. S., Carey, F., and Crawford, N.,** Characterization of the calcium-sequestering process associated with human platelet intracellular membranes isolated by free-flow electrophoresis, *Biochem. J.,* 222, 413, 1984.

62. **Imaoka, T., Lynham, J. A., and Haslam, R. J.,** Purification and characterization of the 47000 dalton protein phosphorylated during degranulation of human platelets, *J. Biol. Chem.,* 258, 11404, 1983.

63. **Shaw, J. O. and Lyons, R. M.,** Requirements for different Ca^{2+} pools in the activation of rabbit platelets. I. Release reaction and protein phosphorylation, *Biochim. Biophys. Acta,* 714, 492, 1982.

64. **Daniel, J. L., Molish, I. R., and Holmsen, H.,** Myosin phosphorylation in intact platelets, *J. Biol. Chem.,* 256, 7510, 1981.

65. **Peleg, I., Muhlrad, A., Eldor, A., Groschel-Stewart, U., and Kahane, I.,** Characterization of the ATPase activities of myosins isolated from the membrane and the cytoplasmic fractions of human platelets, *Arch. Biochem. Biophys.,* 234, 442, 1984.

66. **Verhoeven, A. J. M., Gorter, G., Mommersteeg, M. E., and Akkerman, J. W. N.,** The energetics of early platelet responses. Energy consumption during shape change and aggregation with special reference to protein phosphorylation and the polyphosphoinositide cycle, *Biochem. J.,* 228, 451, 1985.

67. **Costa, J. L., Reese, T. S., and Murphy, D. L.,** Serotonin storage in platelets: estimation of storage-packet size, *Science,* 183, 537, 1974.

68. **Holmsen, H.,** Mechanisms of platelet secretion, in *Platelets: Cellular Response Mechanism and Their Biological Significance,* Rotman, A., Ed., John Wiley & Sons, New York, 1980, 150.

69. **Holmsen, H. and Day, H. J.,** The selectivity of the thrombin-induced platelet release reaction. Subcellular localization of released and retained substances, *J. Lab. Clin. Med.,* 75, 840, 1970.

70. **Detwiler, T. C. and Feinman, R. D.,** Kinetics of the thrombin-induced release of adenosine triphosphate by platelets. Comparison with release of calcium, *Biochemistry,* 12, 2462, 1973.

71. **Robblee, L. S., Kornstein, L. B., and Shepro, D.,** Calcium electrode. A method for continuous monitoring of the platelet release reaction, *Thrombosis Haemostas.,* 37, 407, 1977.

72. **Murer, E. H.,** Use of preabsorbed serotonin for measuring the release reaction, in Platelet Function Testing, Day, H. J., Holmsen, H., and Zucker, M. B., Eds., DHEW Publ. NIH-78-1087, U.S. Department of Health, Education, and Welfare, Washington, D.C., 1978, 142.

73. **Chesney, C. M., Pifer, P. D., Byers, L. W., and Muirhead, E. E.,** Effect of Platelet Activating Factor (PAF) on human platelets, *Blood,* 59, 582, 1982.

74. **Zucker, M. B., Mosesson, M. W., Broekman, M. J., and Kaplan, K. L.,** Release of platelet fibronectin, cold-insoluble globulin from α-granules induced by thrombin or collagen, *Blood,* 54, 8, 1979.

75. **Koutts, J., Walsh, P. N., Plow, E. F., Fenton, J. W., Bouma, B. N., and Zimmerman, T. S.,** Active release of human platelet factor VIII-related antigen by adenosine diphosphate, collagen and thrombin, *J. Clin. Invest.,* 62, 1255, 1978.

Index

INDEX

A

2A23187, 7, 54, 97, 138, 139
ABP, see Actin-binding protein
4-Acetamido-4′-isothiocyanostilbene 2,2′-disulfonic
    acid (SITS), 52, 53, 55
Acetylcholine, 114
Acid oxidation, 113
Acinar cells, 4
Actin, 84—86, 96, 128
Actin-binding protein (ABP), 88
α-Actinin, 89
β-Actinin, 88
Actin-treadmilling, 128
Activated cells, 4, see also specific types
Actomyosin system, 81—99, see also specific
        components
  actin, see Actin
  actin-binding protein, 88,
  acumentin, 88
  ankyrin, 90
  caldesmon, 88—89
  calmodulin, 83, 84
  contractile proteins, 92—99
  cytochalasins, 90
  DNAase I, 87
  gelsolin, 87—88
  myosin, 82—83
  myosin-associated proteins, 82—84
  myosin light chain kinase, 83
  nonmuscle, 90—92
  phalloidin, 90
  phosphatase, 83—84
  phosphorylation and, 91—92
  platelet, 91—92
  profilin, 87
  protein kinase C, 84
  spectrin, 90
  talin, 89—90
  tropomyosin, 86—87
  vinculin, 89
Acumentin, 88
Adenosine diphosphate (ADP), 128
Adenosine monophosphate (AMP), cyclic, see Cyclic
    AMP
Adenosine triphosphatase (ATPase), 48—50, 53,
    90—92
Adenosine triphosphate (ATP), 6, 7, 110, 116
  catecholamine secretion and, 57
  in chromaffin granules, 114
  cytosolic, 39, 115
  demand for, 130—131
  energy metabolism and, 140—142
  enzymes utilizing, 111
  in exocrine pancreas, 114
  free enthalpy for hydrolysis of, 117
  homeostasis of, 127—128

  hydrolysis of, 117, 134—137
  magnesium-dependent, 56
  in mast cells, 114
  in pancreatic islets, 116, 118
  potassium channels and, 35, 39, 118
  production of, 113
  rapid blockade of resynthesis of, 132—134
  resynthesizing pathways of, 126—130
  supply of, 130—131
  utilization of, 113, 116
Adenylate cyclase, 8, 84
Adenylate energy charge, 113, 116, 117, 128
ADP, see Adenosine diphosphate
Adrenal chromaffin cells, 4, 5
Adrenal medullary chromaffin cells, 48, 93—94, 114
Adrenal medullary pancreatic cells, 131
Aequorin, 6
Aldolase, 127, 130
Amiloride, 7
Amino acids, 114, see also specific types
2-Aminobicyclo-(2.2.1)heptane-2-carboxylic acid, 39
AMP, see Adenosine monophosphate
Anaphylactic reaction, 113
ANF, see Atrial natriuretic factor
Anion channel blocking drugs, 54, see also specific
      types
Anion-sensitive ATPase, 49
Anion transport, 51
Ankyrin, 90
Anterior pituitary cells, 5
Antimicrotubule agents, 97, see also specific types
ATP, see Adenosine triphosphate
Atrial natriuretic factor (ANF), 9

B

Baby hamster kidney cells, 83
Band III protein, 52
Basophils, 55
B-cells, pancreatic, see Pancreatic B-cells
BCNU, see 1,3-Bis(2-chloroethyl)-1-nitrosourea
Bilayer dehydration, 72
Bilayer lipid mixing, 66
1,3-Bis(2-chloroethyl)-1-nitrosourea (BCNU), 36
Brain cells, 83, 87

C

Calcium
  accumulation of, 8
  in activated cells, 4
  catecholamine secretion dependent on, 57
  gelsolin and, 88
  homeostasis of, 4
  increase in, 7
  insulin secretion dependent on, 57
  low concentrations of, 56

D

E

Milton Keynes UK
Ingram Content Group UK Ltd.
UKHW051935141024
449569UK00027B/1496